Spatial Grasp as a Model for Space-based Control and Management Systems

Governmental agencies and private companies of different countries are actively moving into space around Earth with the aim to provide smart communication and industry, security, and defense solutions. This often involves massive launches of small, cheap satellites in low earth orbits, which is also contributing to the growth of space debris. The book offers a high-level holistic system philosophy, model, and technology that can effectively organize distributed space-based systems, starting with their planning, creation, and growth. The Spatial Grasp Technology described in the book, based on parallel navigation and pattern-matching of distributed environments with high-level recursive mobile code, can effectively provide any networking protocols and important system applications, by integrating and tasking available terrestrial and celestial equipment. This book contains practical examples of technology-based solutions for tracing hypersonic gliders, continuing observation of certain objects and infrastructures on Earth from space, space-based command and control of large distributed systems, as well as collective removal of increasing amounts of space junk. Earlier versions of this technology were prototyped and used in different countries, with the current version capable of being quickly implemented in traditional industrial or even university environments. This book is oriented toward system scientists, application programmers, industry managers, and university students interested in advanced MSc and PhD projects related to space conquest and distributed system management.

Dr Peter Simon Sapaty, Chief Research Scientist, Ukrainian Academy of Sciences, has worked with networked systems for five decades. Outside of Ukraine, he has worked in the former Czechoslovakia (now Czech Republic and Slovakia), Germany, the UK, Canada, and Japan as a group leader, Alexander von Humboldt researcher, and invited and visiting professor. He launched and chaired the Special Interest Group (SIG) on Mobile Cooperative Technologies in Distributed Interactive Simulation project in the United States, and invented a distributed control technology that resulted in a European patent and books with Wiley, Springer, and Emerald. He has published more than 250 papers on distributed systems and has been included in the Marquis Who's Who in the World and Cambridge Outstanding Intellectuals of the 21st century. Peter also works with several international scientific journals.

Spatial Grasp as a Model for Space-based Control and Management Systems

Peter Simon Sapaty

CRC Press is an imprint of the
Taylor & Francis Group, an **informa** business

First edition published 2022
by CRC Press
6000 Broken Sound Parkway NW, Suite 300, Boca Raton, FL 33487-2742

and by CRC Press
4 Park Square, Milton Park, Abingdon, Oxon, OX14 4RN

CRC Press is an imprint of Taylor & Francis Group, LLC

ISBN: 978-1-032-13609-7 (hbk)
ISBN: 978-1-032-13610-3 (pbk)
ISBN: 978-1-003-23009-0 (ebk)

DOI: 10.1201/9781003230090

Typeset in Sabon
by SPi Technologies India Pvt Ltd (Straive)

For encouraging and supporting the preparing of this book to my wife Lilia, son Alex, grandsons Eugene and Vanya, and mother-in-law Valentina, 97, our high school geography teacher, who now sees a great opportunity to teach geography from Space.

Contents

Acknowledgments

Naming here the international scientists, also good colleagues for years, who supported this book from the very beginning and on its full way to the completion, and whose invaluable suggestions and critical comments helped much in finalizing its orientation and contents.

—John Page, University of New South Wales (UNSW), Sydney, Australia, who also supported my previous publications, including *Journal of Intelligent Unmanned Systems*, Emerald Publishing, edited by him, where we cooperate on its advisory board, and the subsequent books by Emerald and Springer.

—Stephen Lambacher, School of Social Informatics, Aoyama Gakuin University, Tokyo, who successfully combined American and Japanese cultures, and with whom we worked at the University of Oita in Japan and discussed hot problems on Earth and in space, and who reviewed the previous books too.

—Bob Nugent (CDR, USN Retired), now with the Busch School of Business, Catholic University of America, Washington, D.C., with whom we cooperated in Japan within international conferences on Artificial Life and Robotics and afterwards on System Management, and who supported previous books too.

—Lubomir Bic, Department of Computer Science, University of California, Irvine, who supported the previous technology version called WAVE demonstrated at UCI many years ago, and whose comment on its spatial nature like "being in the driver's seat" remains very useful for the new versions too.

Active communications and discussions with different scientific journals and their highly professional staff, who organized urgent invited publications of key ideas related to the current book, were extremely helpful and productive. With special thanks to:

—Svitlana Tymchyk, Nataliia Karevina, and Yevheniia Petruk, *Mathematical Machines and Systems journal* (ISSN 1028-9763), of the National Academy of Sciences of Ukraine. http://www.immsp.kiev.ua/publications/eng/index.html

—Siva Deepthi, Acta Scientific Computer Sciences, where Acta Scientific is an open access publishing group oriented toward intellectualizing the global society and advancements in scientific research. https://www.actascientific.com/ASCS.php

—Bruce Lee, Network and Communication Technologies, Canadian Center of Science and Education, ISSN(Print): 1927-064X, ISSN(Online): 1927-0658. https://www.ccsenet.org/journal/index.php/nct

Preface

Different countries are chaotically rushing into space, actually Klondike-like adventures, in hope to provide quick and smart communication, industrial, security, and defense solutions. And this massive, often competitive, penetration into space in an attempt to gain something quickly and cheaply is not based on any global culture and planning. It often involves massive launches of simple and unsafe small satellites in Low Earth Orbits (LEOs), which are also contributing to the growth of space debris endangering any future space research and conquest. We witnessed a similar chaotic situation more than half a century ago, but on Earth, being actively engaged in creation of first citywide heterogeneous computer networks, well before the internet, with very different machines and absence of communication channels between them. And we are eager now to use this and the subsequent experience, accumulated in different countries, for the creation and management of rapidly growing celestial systems, for which a sort of global philosophy and smart technological solutions may be of particular importance.

The first linking of our work on distributed networked systems with space was in 1987 when, working at the Basic International AI lab in Bratislava (now Slovak Republic), we met at Frontiers in Computing conference in Amsterdam, a U.S. team presenting their work on Strategic Defense Initiative (SDI) based on multiple small satellites called "brilliant pebbles." And the idea was born to connect our research on distributed computer and knowledge networks in the language called WAVE with organization of space systems too. It strengthened after recent announcement by Space Development Agency (SDA) of its multilayered Space Architecture oriented toward numerous cooperating satellites, mostly on LEOs. This project appeared to be a perfect match to the developed Spatial Grasp Model and Technology, where cooperative work of distributed dynamic systems is one of the key orientations and applications.

This book offers a unified approach to organization of large multiple satellite systems, often called meta-constellations, on their full way from initial planning, creation, and networking to growth and maturity, using a special high-level model and technology based on parallel wavelike self-spreading spatial scenarios, which, in their infancy, proved fruitful for the creation of first computer networks. It contains many examples of technology-based practical solutions like creation and management of the basic transport layer for satellite constellations, effective tracing of complexly moving objects like hypersonic gliders, continuing observation of important objects and their infrastructures on Earth from space, space-based command and control of large distributed systems and operations, and also collective removal of growing space junk using special cleaning constellations. Earlier versions of the technology, described in six previous books, were prototyped and used in different countries, and the current version can also be implemented by a group of system programmers even in traditional university environments.

In conclusion, we would like to stress that this book has purely scientific and technological orientation, and the mentioned concrete multiple satellite projects were only used to demonstrate potential applicability of the developed spatial model and technology for dealing with complex celestial systems. The obtained results can be applied in many other areas too, including space economy, communications, robotics, weather prediction, etc., with related examples also described in the previous books.

Kiev, Ukraine
November 2021
Peter Simon Sapaty

Chapter 1

Introduction

1.1 THE RUSH INTO SPACE, EXISTING PROBLEMS, AND SOLUTIONS NEEDED

Humanity's interest in the heavens has been universal and enduring. Human space exploration helps us address fundamental questions about our place in the universe and the history of our solar system. Through addressing the challenges related to human space exploration, we expand technology, create new industries, and foster peaceful connection between nations. Curiosity and exploration are vital to the human spirit, and the challenge of going deeper into space will invite the citizens of the world today and the generations of tomorrow to unite on this exciting journey. We can expect space solar power plants, industrial exploration of the Moon, eco-industry, recovery of natural resources, global weather management, large-scale artificial structures in space, the use of raw materials of other planets, colonization of Venus and Mars, etc., including the already evolving space industry and robotics [1–18].

Different countries are often chaotically rushing into space in hope to provide quick and smart communication and industrial, security, and defense solutions. And this massive, often competitive, penetration into space in an attempt to gain something quickly and cheaply is not based on any global culture and planning, say UN supported. It often involves massive launches of cheap and unsafe small satellites in low Earth orbits (LEOs), which are also contributing to the growth of space debris endangering any future space research and conquest. In total, 11,139 satellites have been launched, out of which 7,389 were in space by the end of April 2021, while the rest have either been burnt up in the atmosphere or have returned to Earth in the form of debris. According to NASA [2], there are millions of pieces of junk flying in LEO, which comprises of spacecraft, tiny flecks of paint from spacecraft, parts of rockets and satellites that are either dead or lost, including objects that are results of explosions in the space. Efficient management of numerous space projects and problems, including those related to global defense like the New Space Architecture of Space Development Agency (SDA) [14–18] and the rapidly growing space debris [19–34], needs serious investigation and development in scientific and technological areas.

DOI: 10.1201/9781003230090-1

1.2 SOME HISTORY OF DEALING WITH LARGE DISTRIBUTED SYSTEMS

The current author witnessed a similar situation more than half a century ago, but on Earth, being actively engaged in the creation of first heterogeneous computer networks combining very different and even incompatible machines, well before the internet. Having offered at that time how to unite such distributed equipment into capable complexes using a sort of parallel wavelike (even virus-like) model, the author is eager now to use a conceptually similar model—spatial grasp model. This model is much more advanced and tested in different countries and the resulting networking technology brings certain order into this enormously growing and often uncontrollable multiple satellite and junk mess around the Earth.

This book actually inherits practical works on creation of citywide computer networks in Kiev, Ukraine, from the end of 1960s, which were integrating different institutes of the National Academy of Sciences and other organizations [35–37]. By spreading a fully interpreted scenario code in a wavelike mode between different computers, we were able to solve complex analytic-numerical problems on heterogeneous computer networks that were difficult to solve using individual computers. These works resulted in a new management concept and distributed control methodology and technology which were further developed in different countries (including former Czechoslovakia, Germany, United Kingdom, United States, Canada, and Japan), with applications in such areas as intelligent network management, industry, social systems, psychology, collective robotics, security, and defense. The technology-based international projects were supported by Siemens and Alexander von Humboldt Foundation in Germany, Ericsson and Defense Research Agency in the United Kingdom, Japan Society for the Promotion of Science, and Distributed Interactive Simulation project in the United States which hosted Special Interest Group on Mobile Cooperative Technologies chaired by the author. The developed concept was demonstrated at the Universities of Braunschweig and Karlsruhe in Germany, Oxford and Surrey in the United Kingdom, British Columbia in Canada, Oita and Aizu in Japan, and California at Irvine in the United States. A number of successful implementations have been made of this approach in such programming languages as Analytic, Fortran, Lisp, and C.

1.3 NEW PHILOSOPHY, MODEL, AND TECHNOLOGY FOR THE MANAGEMENT OF SPACE

A special high-level recursive Spatial Grasp Language (SGL) has been developed in which distributed, parallel, and holistic algorithms could be expressed with resultant spatial scenarios in a much more compact and simpler way than in other languages. All this activity resulted in a European patent and more than 200 international publications, including six books of Wiley, Springer, and Emerald [38–108]. The aim of the current book is to generalize all these works and obtained experience in the form of a new computational, control, and management model as a natural extension of traditional concept of algorithm, with orientation on applications in very large distributed systems operating in combined terrestrial and celestial environments. This model allows us to express complex solutions in distributed spaces with feeling of direct staying and moving in and through them, also to obtain their overall vision and understanding in a holistic manner and organize distributed space-based systems on different stages of their development and growth. The resultant Spatial Grasp Technology (SGT), based on parallel conquest and pattern-matching of distributed environments with high-level recursive mobile code, can effectively provide any networking protocols and important applications of large satellite constellations, especially those in LEOs.

The book contains examples of technology-based solutions for establishing basic communications between satellites, starting from their initial, often chaotic, launches and distributing and collecting data in the growing constellations with even unstable and rapidly changing connections between satellites. It describes how to organize and register networking topologies in case of predictable distances between satellites, and how the fixed networking structures can help in solving complex problems. The latter including those related to the SDA multiple-satellite architecture and allowing for effective integration of its continuous Earth observation and missile tracking layers based on self-spreading mobile intelligence, and also fighting enormously growing space debris by collectively behaving constellations of cleaning satellites.

What is shown in Figure 1.1 may be considered as a symbolic expression, or formula, of the main idea of the book, which combines areas of investigation (on both sides) with the spatial grasp model for their management and control (centered).

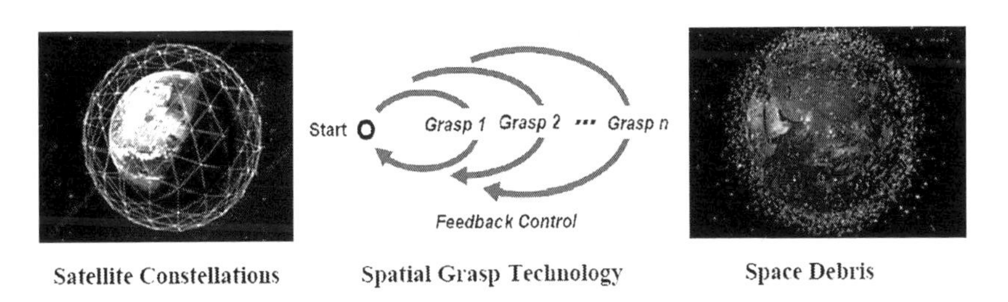

Figure 1.1 A symbolic formula of the book.

1.4 SUMMARY OF BOOK CHAPTERS

1.4.1 Chapter 2: Satellite constellations, projects, and debris

It provides a summary of existing publications on satellite constellations and mega-constellations with their features and problems. These include proper management, using onboard intelligence, providing replacement, issues of effective communications which may be optical, complexity of ground antennas and gateways to monitor low orbit satellites, and also the increase of space debris by constellation units if not de-orbited at the end of service or after failure. The chapter briefs some defense projects with multiple satellites in space such as Strategic Defense Initiative of the eighties, and the most recent SDA's Next-Generation Space Architecture. Commercial activity and projects are reviewed too. A summary on the debris problems and existing solutions is provided with a mention on legal questions of junk removal, debris surveillance and tracking, already existing removal contracts and techniques, as well as first removal missions. It also stresses the need for a unified system approach for dealing with different types of satellites constellations, especially in LEO orbits.

1.4.2 Chapter 3: Spatial Grasp Model (SGM) and Spatial Grasp Technology (SGT)

It briefs the traditional concept of algorithm as a finite sequence of well-defined, computer-implementable instructions, and widely used flowchart as a diagram representing workflow or process. The chapter describes basics of Spatial Grasp (SG) model and how it differs from conventional algorithm; it also introduces a new type of a chart, called spatiochart, for describing scenarios operating directly in distributed spaces. It shows how different collections of actions can be described in SG and exhibited by spatiocharts, including those using control rules, which are supervising repetition, sequencing and branching in spatial scenarios, also expressing spatial dataflow and exchange, with such organizations being potentially hierarchical and recursive. It describes main elements of Spatial Grasp Technology (SGT) and its high-level recursive Spatial Grasp Language (SGL) based on the SG philosophy. This includes different types of worlds SGT operates with, various constants which may represent information or physical matter, repertoire of spatial variables of SGL which may be stationary or mobile, main types of SGL rules that can be arbitrarily nested, different control states provided by SGL scenarios propagation, as well as general organization of the distributed and networked SGL interpreter.

1.4.3 Chapter 4: Spatial Grasp Language (SGL)

The chapter briefs main concepts of SGT established in the previous chapter and then provides full SGL syntax description with peculiarities of its recursive organization. It describes SGL constants which include information, physical matter, special and custom constants, as well as arbitrary complex or compound ones. It also gives the repertoire of SGL spatial variables, which include global, heritable, frontal, nodal, and environmental variables which can be stationary or mobile. It describes and explains main SGL rules which express and guarantee such functionalities and operations as: usage, movement, creation, echoing, verification, assignment, advancement, branching, transference, exchange, timing, qualification, and grasping. After the detailed SGL description, the chapter presents and explains some examples of spatial scenarios in SGL which include: distributed network management, organization of integral human-robotic collectives, and simulation of spreading of malicious viruses and fighting with them by worldwide distribution of antivirus vaccines.

1.4.4 Chapter 5: Elementary constellation operations under SGT

After a brief summary of SGT described in detail in Chapters 3 and 4, the current chapter describes how to integrate satellite constellations into a capable system under SGT, and explains how to provide broadcasting of executive orders to all satellites from a ground station. It then shows how supplying satellites with SGL interpreters which can communicate directly with each other as an integral system and also with the ground stations may convert the constellation into a self-organized and entirely space-located system, with significant simplification of ground antennas and reduction of their numbers. The chapter demonstrates some basic operations over satellite constellations in SGL in a virus-like self-spreading parallel mode, which includes broadcasting executive orders to all satellites via direct and changeable communications between them, collecting and returning accumulated data by all satellites, and constellation repositioning and restructuring which may be emergent. It also mentions more complex and advanced constellation solutions that can be provided under SGT, which are discussed in the subsequent chapters.

1.4.5 Chapter 6: Transport layer organization under SGT

It first reminds a brief summary of SGT described in detail in Chapters 3 and 4, then briefs on SDA Transport Layer aimed at providing basic satellite communications and networking for the new space architecture. The chapter shows how to deal with highly dynamic constellation topologies under SGT, which includes finding proper satellite nodes, which may be arbitrarily remote, and delivery of a package to all of them, reaching proper destinations with the return and collection of certain items from them all, also introducing special measures for dealing with high unpredictability of topology dynamics. After this, it demonstrates how to work with stable constellation topologies while explaining advantages of them, shows how to create and fix communication topology in case of stable distances between satellites, and how to express in SGL basic networking operations like finding shortest paths and routing tables. Other demonstrations include discovery of certain structures and components in distributed networks with stable topologies, like articulation points and cliques. At the end, a routine satellite constellation management example is shown under stable topologies like collection and return of remote data, and how trivial this can be in the network stability case.

1.4.6 Chapter 7: Advanced space projects management under SGT

After a brief summary of SGT described in detail in Chapters 3 and 4, it first gives an example of tasking SDI brilliant pebbles in SGL, which would mostly operate on their own with minimum or even without communications with others pebbles. It then shows how self-spreading SGL scenarios within the SDA Architecture Tracking Layer can effectively follow and destroy complexly moving objects like hypersonic gliders in arbitrary large distributed spaces with intensive cooperation, actually team work, of different satellites, also shows how to organize multithreaded tracking for increased safety and in case of unstable satellite networks. The chapter also demonstrates how to use mobile SGL code for implementation of SDA Custody Layer oriented on constant monitoring of terrestrial objects of particular interest. At the end, it shows how to organize highly cooperative work, actually integration, of custody and tracking layers under stable constellation topologies.

1.4.7 Chapter 8: Using virtual layer for constellation management

The chapter starts from a brief summary of SGT described in detail in Chapters 3 and 4. It then introduces a virtual layer for the custody support, where custody-related virtual nodes reflecting objects on Earth are constantly managed and updated regardless of satellites' movement is space. The chapter provides examples of control of custody operations via the virtual layer, including verifying changing distances between remote custody locations. It uses regularly updated virtual layer for worldwide delivery of goods, also analyzing complex distributed system as a whole on the example of forest fires covering different regions with different fire intensities. The chapter describes distributed virtual-physical command and control solution with regularly updated correspondence between virtual and physical Earth-based command and control, also delivering command orders via virtual layer and managing two- and then multilevel virtual-physical C2 infrastructure. Such symbiosis may allow us to effectively supervise terrestrial, celestial, and combined missions even completely from space, say, in emergency situations on Earth.

1.4.8 Chapter 9: Space debris removal under SGT

The chapter first repeats a brief summary of SGT described in detail in Chapters 3 and 4. It then shows how a special constellation of junk-cleaning satellites can be converted under SGT into an intelligent team capable of organizing and executing massive debris removal operations with a good deal of autonomy and reduced ground communications. It demonstrates in SGL how the debris discovery and organization-execution scenarios can be expressed in SGL which is interpreted in a distributed way by the whole satellite cleaning network, where the removal solutions can be organized simultaneously for multiple debris. The chapter also provides a solution in SGL where the junk items can be virtually treated as active objects traveling around the Earth and finding proper cleaners on their own initiative, which guarantees any time needed for finding a suitable match in such self-removal procedure. At the end, an SGL solution is shown for effective integration of different junk-removal strategies.

1.4.9 Chapter 10: Conclusions

Traditional parallel and distributed computing, control, and management models and approaches suppose that there already exist systems having identifiable components which are somehow interacting with each other, also forming together a sort of a more or less stable structure. But on the current chaotic rush into space, at least at the beginning, there may be situations where the multiple launched objects may not represent any system at all. But the described spatial model and technology can solve different problems by runtime and holistic grasping of multitudes of such initially disorganized objects and converting them into operable systems, actually on the fly. As was shown throughout the book, SGT can also effectively integrate these space objects into stable systems and networks capable of solving much more complex problems, actually covering the whole space-conquering process from its birth to maturity under the same universal spatial model and language. The described technology can be readily implemented in existing industrial and even standard university environments, as was successfully done in different countries for its previous versions.

REFERENCES

1. N. Mohanta, How many satellites are orbiting the Earth in 2021? Geospatial World, May 28, 2021. https://www.geospatialworld.net/blogs/how-many-satellites-are-orbiting-the-earth-in-2021/
2. Space Debris, NASA Headquarters Library, https://www.nasa.gov/centers/hq/library/find/bibliographies/space_debris
3. United Nations Register of Objects Launched into Outer Space, The United Nations Office for Outer Space Affairs. http://www.unoosa.org/oosa/en/spaceobjectregister/index.html

4. G. Martin, NewSpace: The «Emerging» Commercial Space Industry. https://ntrs.nasa.gov/archive/nasa/casi.ntrs.nasa.gov/20140011156.pdf
5. J.-M. Bockel, The Future of the Space Industry. General Report. November 17, 2018. https://www.nato-pa.int/download-file?filename=sites/default/files/2018-12/2018%20-%20THE%20FUTURE%20OF%20SPACE%20INDUSTRY%20-%20BOCKEL%20REPORT%20-%20173%20ESC%2018%20E%20fin.pdf
6. G. Curzi, D. Modenini, P. Tortora, Review large constellations of small satellites: A survey of near future challenges and missions, *Aerospace* 7(9) (2020), 133. https://www.mdpi.com/2226-4310/7/9/133/htm
7. A. Venkatesan, J. Lowenthal, P. Prem, M. Vidaurri, The impact of satellite constellations on space as an ancestral global commons, *Nature Astronomy* 4 (2020, November), 1043–1048. www.nature.com/natureastronomy
8. R. Skibba, How satellite mega-constellations will change the way we use space, *MIT Technology Review*, February 26, 2020. https://www.technologyreview.com/2020/02/26/905733/satellite-mega-constellations-change-the-way-we-use-space-moon-mars/
9. M. Minet, The space legal issues with Megs-Constelllations, November 3, 2020. https://www.spacelegalissues.com/mega-constellations-a-gordian-knot/
10. E. Siegel, Astronomy faces a mega-Crisis as satellite mega-constellations loom, January 19, 2021. https://www.forbes.com/sites/startswithabang/2021/01/19/astronomy-faces-a-mega-crisis-as-satellite-mega-constellations-loom/?sh=30597dca300d
11. A. Jones, China is developing plans for a 13,000-satellite megaconstellation, *Space News*, April 21, 2021. https://spacenews.com/china-is-developing-plans-for-a-13000-satellite-communications-megaconstellation/
12. N. Reilanda, A. J. Rosengren, R. Malhotra, C. Bombardelli, Assessing and minimizing collisions in satellite mega-constellations, 2021, Published by Elsevier B.V. on behalf of COSPAR. https://www.sciencedirect.com/science/article/abs/pii/S0273117721000326
13. J. Fous, Mega-constellations and mega-debris, October 10, 2016. https://www.thespacereview.com/article/3078/1
14. Space Development Agency Next-Generation Space Architecture. 2019. https://www.airforcemag.com/PDF/DocumentFile/Documents/2019/SDA_Next_Generation_Space_Architecture_RFI%20(1).pdf
15. S. Magnuson, Web exclusive: Details of the Pentagon's new space architecture revealed. 2019. https://www.nationaldefensemagazine.org/articles/2019/9/19/details-of-the-pentagon-new-space-architecture-revealed.
16. Messier D. Space Development Agency seeks next-gen architecture in first RFI. 2019. http://www.parabolicarc.com/2019/07/07/space-development-agency-issues-rfi/
17. V. Insinna, Space agency has an ambitious plan to launch 'hundreds' of small satellites. Can it get off the ground? *DefenceNews*. Space. April 10, 2019. https://www.defensenews.com/space/2019/04/10/sda-has-an-ambitious-plan-to-launch-hundreds-of-small-satellites-can-it-get-off-the-ground/
18. N. Strout, Space Development Agency approves design for satellites that can track hypersonic weapons, C4ISRNET, 2021. https://www.c4isrnet.com/battlefield-tech/space/2021/09/20/space-development-agency-approves-design-for-satellites-that-can-track-hypersonic-weapons/

19. Research on space debris, safety of space objects with nuclear power sources on board and problems relating to their collision with space debris. *Committee on the Peaceful Uses of Outer Space*. Vienna (2019). http://www.unoosa.org/res/oosadoc/data/documents/2019/aac_105c_12019crp/aac_105c_12019crp_7_0_html/AC105_C1_2019_CRP07E.pdf
20. Deorbit Systems, National Aeronautics and Space Administration, November 28, 2020. https://www.nasa.gov/smallsat-institute/sst-soa-2020/passive-deorbit-systems
21. A. Froehlich (Ed.), *Space Security and Legal Aspects of Active Debris Removal*, Springer (2019). https://www.springer.com/gp/book/9783319903378
22. A. Sheer, S. Li, Space debris mounting global menace egal issues pertaining to space debris removal: Ought to revamp existing space law regime, *Beijing Law Review*, 10 (2019), 423–440. https://www.scirp.org/pdf/BLR_2019051615104007.pdf
23. G. S. Aglietti, From space debris to NEO, some of the major challenges for the space sector, *Frontiers in Space Technologies* 16 June 2020. https://www.frontiersin.org/articles/10.3389/frspt.2020.00002/full
24. L. David, Space junk removal is not going smoothly, *Scientific American*, April 14, 2021. https://www.scientificamerican.com/article/space-junk-removal-is-not-going-smoothly/
25. ESA commissions world's first space debris removal, *ESA / Safety & Security / Clean Space*, December 9, 2019. https://www.esa.int/Safety_Security/Clean_Space/ESA_commissions_world_s_first_space_debris_removal
26. A. Parsonson, ESA signs contract for first space debris removal mission, *Space News*, December 2, 2020. https://spacenews.com/clearspace-contract-signed/
27. Humza, The first-ever space mission to clean orbital junk will use a giant claw, *Techspot*, December 1, 2020. https://www.techspot.com/community/topics/the-first-ever-space-mission-to-clean-orbital-junk-will-use-a-giant-claw.266509/
28. Japanese company planning space debris removal by laser on satellite, *Kyodo News*, August 8, 2020. https://english.kyodonews.net/news/2020/08/fc06829d1d9a-japanese-company-planning-space-debris-removal-by-laser-on-satellite.html
29. Plasma thruster: New space debris removal technology, *Tohoku University*, September 27, 2018. https://www.eurekalert.org/pub_releases/2018-09/tu-ptn092718.php
30. K. Hunt, Mission to clean up space junk with magnets set for launch, *CNN*, April 1, 2021. https://edition.cnn.com/2021/03/19/business/space-junk-mission-astroscale-scn/index.html
31. M. Obe, Japan's Astroscale launches space debris-removal satellite, *Nikkei Asia*, March 22, 2021. https://asia.nikkei.com/Business/Aerospace-Defense/Japan-s-Astroscale-launches-space-debris-removal-satellite
32. C. Weiner, New effort to clean up space junk reaches orbit, March 21, 2021. https://www.npr.org/2021/03/21/979815691/new-effort-to-clean-up-space-junk-prepares-to-launch
33. Y. Chen et al., Optimal mission planning of active space debris removal based on genetic algorithm, *IOP Conf. Series: Materials Science and Engineering* 715 (2020), 012025. https://iopscience.iop.org/article/10.1088/1757-899X/715/1/012025/pdf

34. R. Klima et al., Space debris removal: Learning to cooperate and the price of anarchy, *Frontiers in Robotics and AI*, 4 June 2018. https://www.frontiersin.org/articles/10.3389/frobt.2018.00054/full
35. A.T. Bondarenko, S.B. Mikhalevich, A.I. Nikitin, P.S. Sapaty, "Software of BESM-6 computer for communication with peripheral computers via telephone channels," in *Computer Software*, Vol. 5, Inst. of Cybernetics Press, Kiev, 1970 (in Russian).
36. A.T. Bondarenko, V.P. Karpus, S.B. Mikhalevich, A.I. Nikitin, P.S. Sapaty, "Information-computing system ABONENT", Tech. Report No. B178338, All-Union Scientific and Technical Inform. Centre, Moscow, 1972 (in Russian).
37. P.S. Sapaty, "A Method of organization of an intercomputer dialogue in the radial computer systems," in The Design of Software and Hardware for Automatic Control Systems, Inst. of Cybernetics Press, Kiev, 1973 (in Russian).
38. P.S. Sapaty, The WAVE-0 language as a framework of navigational structures for knowledge bases using semantic networks, in *Proceedings of USSR Academy of Sciences*. Technical Cybernetics, No. 5 (1986) (in Russian).
39. P.S. Sapaty, The wave approach to distributed processing of graphs and networks, in *Proceedings of International Working Conference Knowledge and Vision Processing Systems*, Smolenice, November 1986.
40. P.S. Sapaty, A wave language for parallel processing of semantic networks. *Computer Artificial Intelligence* 5(4) (1986), 289–314.
41. P.S. Sapaty, S. Varbanov, M. Dimitrova, Information systems based on the wave navigation techniques and their implementation on parallel computers, in *Proceedings of International Working Conference on Knowledge and Vision Processing Systems*, Smolenice, November 1986.
42. P.S. Sapaty, I. Kocis, A parallel network wave machine, in *Proceedings of 3rd International Workshop PARCELLA'86* (Akademie-Verlag, Berlin, 1986).
43. P. Sapaty, S. Varbanov, A. Iljenko, The WAVE model and architecture for knowledge processing, in *Proceedings of Fourth International Conference on Artificial Intelligence and Information-Control Systems of Robots*, Smolenice (1987).
44. P.S. Sapaty, Distributed Artificial Brain for Collectively Behaving Mobile Robots, in *Proceedings of Symposium and Exhibition Unmanned Systems 2001*, Baltimore, MD, 31 July–2 August 2001, 18 p.
45. P.S. Sapaty, A. Morozov, M. Sugisaka, DEW in a Network Enabled Environment, in *Proceedings of the International Conference Directed Energy Weapons 2007*, Le Meridien Piccadilly, London, UK, 28 February–1 March 2007.
46. P.S. Sapaty, Global Management of Distributed EW-Related System, in *Proceedings of International Conference Electronic Warfare: Operations & Systems 2007*, Thistle.
47. P.S. Sapaty, M. Sugisaka, J. Filipe, Making Sensor Networks Intelligent, in *Proceedings of the 4th International Conference on Informatics in Control, Automation and Robotics, ICINCO-2007*, Angers, France, 9–12, May 2007.
48. P.S. Sapaty, Crisis management with distributed processing technology. *International Transactions on Systems Science and Applications* 1(1) (2006), 81–92. ISSN 1751-1461
49. P.S. Sapaty, M. Sugisaka, R. Finkelstein, J. Delgado-Frias, N. Mirenkov, Emergent Societies: An Advanced IT Support of Crisis Relief Missions, in *Proceedings*

of Eleventh International Symposium on Artificial Life and Robotics (AROB 11th'06), Beppu, Japan, 23–26, January 2006, ISBN 4-9902880-0-9
50. P.S. Sapaty, Distributed Technology for Global Dominance, in *Proceedings of International Conference Defense Transformation and Net-Centric Systems 2008, as part of the SPIE Defense and Security Symposium*, World Center Marriott Resort and Convention Center, Orlando, FL, USA, 16–20, March 2008 (Proceedings of SPIE—Volume 6981, Defense Transformation and Net-Centric Systems 2008, Raja Suresh, Editor, 69810T, April 3, 2008).
51. P.S. Sapaty, Distributed Technology for Global Dominance. Keynote lecture, in *Proceedings of the Fifth International Conference in Control, Automation and Robotics ICINCO 2008, The Conference Proceedings, Funchal, Madeira*, Portugal, 11–15 May 2008.
52. P.S. Sapaty, Human-Robotic Teaming: A Compromised Solution. AUVSI's Unmanned Systems North America 2008, San Diego, USA, 10–12 June 2008.
53. P.S. Sapaty, M. Sugisaka, M.J. Delgado-Frias, J. Filipe, N. Mirenkov, Intelligent management of distributed dynamic sensor networks, *Artificial Life and Robotics* 12(1–2) (2008, March), 51–59.
54. P.S. Sapaty, High-Level Communication Protocol for Dynamically Networked Battlefields, in *Proceedings of International Conference Tactical Communications 2009 (Situational Awareness and Operational Effectiveness in the Last Tactical Mile)*, One Whitehall Place, Whitehall Suite and Reception, London, UK (2009).
55. P.S. Sapaty, Distributed capability for battlespace dominance, in *Electronic warfare 2009 conference and exhibition, Novotel London West Hotel and Conference Center*, London, 14–15, May 2009.
56. P.S. Sapaty, Distributed capability for battlespace dominance, in *Electronic warfare 2009 conference and exhibition*, Novotel London West Hotel and Conference Center, London (2009).
57. P.S. Sapaty, Providing Spatial Integrity for Distributed Unmanned Systems, in *Proceedings of 6th International Conference in Control, Automation and Robotics ICINCO 2009*, Milan, Italy (2009).
58. P.S. Sapaty, Distributed Technology for Global Control. Book chapter, *Lecture Notes in Electrical Engineering, Informatics in Control, Automation and Robotics*, vol. 37 (Springer, Berlin, 2009).
59. P.S. Sapaty, Gestalt-Based Integrity of Distributed Networked Systems. SPIE Europe Security + Defence, bcc Berliner Congress Centre, Berlin, Germany (2009).
60. P.S. Sapaty, Remote Control of Open Groups of Remote Sensors, in *Proceedings of SPIE Europe Security + Defence*, Berlin, Germany (2009).
61. P.S. Sapaty, Tactical Communications in Advanced Systems for Asymmetric Operations, in *Proceedings of Tactical Communications 2010*, CCT Venues, Canary Wharf, London UK, 28–30 April 2010.
62. P.S. Sapaty, High-Level Technology to Manage Distributed Robotized Systems, in *Proceedings of Military Robotics 2010*, Jolly St Ermins, London, UK, 25–27, May 2010.
63. P.S. Sapaty, Emerging Asymmetric Threats, Q&A Session. Tactical Communications 2010, CCT Venues, Canary Wharf, London UK, 28–30, April 2010.

64. P.S. Sapaty, High-level Organisation and Management of Directed Energy Systems, in *Proceedings of Directed Energy Weapons 2010*, CCT, Canary Wharf, London UK, 25–26, March 2010.
65. P.S. Sapaty, Formalizing commander's intent by spatial grasp technology. Accepted paper at the international society of military sciences (ISMS) 2012 annual conference, Kingston, Ontario, Canada, 23–24, October 2012.
66. P.S. Sapaty, Unified Transition to Robotized Armies with Spatial Grasp Technology. International Summit Military Robotics, London, United Kingdom, 12–13 November 2012.
67. P.S. Sapaty, Distributed air and missile defense with spatial grasp technology, *Intelligent Systems, Control and Automation: Science and Engineering* 3(2) (2012), 117–131.
68. P.S. Sapaty, Global electronic dominance, in *12th International Fighter Symposium*, 6th–8th November 2012, Grand Connaught Rooms, London, UK.
69. P.S. Sapaty, Providing global awareness in distributed dynamic environments. International summit ISR, London, 16–18, April 2013.
70. P.S. Sapaty, Ruling distributed dynamic worlds with spatial grasp technology. Tutorial at the international science and information conference 2013 (SAI), London, UK, 7–9, October 2013.
71. P.S. Sapaty, Night vision under advanced spatial intelligence: a key to battlefield dominance. International summit night vision 2013, London, 4–6, June 2013.
72. P.S. Sapaty, Integration of ISR with advanced command and control for critical mission applications. SMi's ISR conference, Holiday Inn Regents Park, London, 7–8, April 2014.
73. P.S. Sapaty, Unified transition to cooperative unmanned systems under spatial grasp paradigm, *International Journals of Transactions and Networks Communication* 2(2) (2014, April).
74. P.S. Sapaty, From manned to smart unmanned systems: a unified transition. SMi's military robotics, *Holiday Inn Regents Park London*, 21–22, May 2014.
75. P.S. Sapaty, Unified transition to cooperative unmanned systems under spatial grasp paradigm, in *19th international command and control research and technology symposium*, Alexandria, Virginia, 16–19, June 2014.
76. P.S. Sapaty, Distributed human terrain operations for solving national and international problems. *International Relations and Diplomacy* 2(9) (2014, September)
77. P.S. Sapaty, Providing over-operability of advanced ISR systems by a high-level networking technology. SMI's airborne ISR, Holiday Inn Kensington Forum, London, United Kingdom, 26–27, October 2015.
78. P.S. Sapaty, Distributed missile defence with spatial grasp technology. SMi's military space, *Holiday Inn Regents Park London*, 4–5, March 2015.
79. P.S. Sapaty, Military robotics: latest trends and spatial grasp solutions, *International Journal of Advanced Research in Artificial Intelligence* 4(4) (2015)
80. P.S. Sapaty, Organization of advanced ISR systems by high-level networking technology. MMC, No 1 (2016).
81. P.S. Sapaty, Towards massively robotized systems under spatial grasp technology, *Journal of Computer Science and Systems Biology* 9(1) (2016).

82. P.S. Sapaty, Towards wholeness and integrity of distributed dynamic systems, *Journal of Computer Science and Systems Biology* 9(3) (2016).
83. P.S. Sapaty, Towards global goal orientation, robustness and integrity of distributed dynamic systems, *Journal of International Relational Diplomacy* 4(6) (2016, June).
84. P.S. Sapaty, Mosaic warfare: from philosohpy to model to solution, *Mathematical Machines and Systems* 3 (2019).
85. P.S. Sapaty, *Symbiosis of Real and Simulated Worlds under Spatial Grasp Technology*. Springer, 2021. 305 p.
86. P.S. Sapaty, *Complexity in International Security: A Holistic Spatial Approach*. Emerald Publishing, 2019. 160 p.
87. P.S. Sapaty, *Holistic Analysis and Management of Distributed Social Systems*. Springer, 2018. 234 p.
88. P.S. Sapaty, *Managing Distributed Dynamic Systems with Spatial Grasp Technology*. Springer, 2017. 284 p.
89. P.S. Sapaty, *Ruling Distributed Dynamic Worlds*. New York: John Wiley & Sons, 2005. 255 p.
90. P.S. Sapaty, *Mobile Processing in Distributed and Open Environments*. New York: John Wiley & Sons, 1999. 410 p.
91. P.S. Sapaty, Global network management under spatial grasp paradigm. *International Robotics & Automation Journal* 6(3) (2020), 134–148. 134. https://medcraveonline.com/IRATJ/IRATJ-06-00212.pdf
92. P.S. Sapaty, Global network management under spatial grasp paradigm, *Global Journal of Researches in Engineering: J General Engineering* 20(5) (2020), 58–81. https://globaljournals.org/GJRE_Volume20/6-Global-Network-Management.pdf
93. P.S. Sapaty, Advanced terrestrial and celestial missions under spatial grasp technology, *Aeronautics and Aerospace Open Access Journal* 4(3) (2020). https://medcraveonline.com/AAOAJ/AAOAJ-04-00110.pdf
94. P.S. Sapaty, Spatial management of distributed social systems, *Journal of Computer Science Research* 2(3) (2020, July). https://ojs.bilpublishing.com/index.php/jcsr/article/view/2077/pdf
95. P.S. Sapaty, Towards global nanosystems under high-level networking technology, *Acta Scientific Computer Sciences* 2(8) (2020). https://www.actascientific.com/ASCS/pdf/ASCS-02-0051.pdf
96. P.S. Sapaty, Symbiosis of distributed simulation and control under spatial grasp technology, SSRG *International Journal of Mobile Computing and Application (IJMCA)* 7(2) (2020, May–August). http://www.internationaljournalssrg.org/IJMCA/2020/Volume7-Issue2/IJMCA-V7I2P101.pdf
97. P.S. Sapaty, Global network management under spatial grasp paradigm, *International Robotics & Automation Journal* 6(3) (2020). https://medcraveonline.com/IRATJ/IRATJ-06-00212.pdf
98. P.S. Sapaty, Global network management under spatial grasp paradigm, *Global Journal of Researches in Engineering: J General Engineering* 20(5) (2020). https://globaljournals.org/GJRE_Volume20/6-Global-Network-Management.pdf
99. P.S. Sapaty, Symbiosis of virtual and physical worlds under spatial grasp technology, *Journal of Computer Science & Systems Biology* 13(6) (2020). https://www.hilarispublisher.com/open-access/symbiosis-of-virtual-and-physical-worlds-under-spatial-grasp-technology.pdf

100. P.S. Sapaty, Symbiosis of real and simulated worlds under global awareness and consciousness, *The Science of Consciousness* | *TSC* 2020. https://eagle.sbs.arizona.edu/sc/report_poster_detail.php?abs=3696
101. P.S. Sapaty, Spatial grasp as a model for space-based control and management systems, *Mathematical Machines and Systems* 1 (2021), 135–138. http://www.immsp.kiev.ua/publications/articles/2021/2021_1/Sapaty_book_1_2021.pdf
102. P.S. Sapaty, Managing multiple satellite architectures by spatial grasp technology, *Mathematical Machines and Systems*, 1 (2021), 3–16. http://www.immsp.kiev.ua/publications/eng/2021_1/
103. P.S. Sapaty, Spatial management of large constellations of small satellites, *Mathematical Machines and Systems* 2 (2021). http://www.immsp.kiev.ua/publications/articles/2021/2021_2/02_21_Sapaty.pdf
104. P.S. Sapaty, Global management of space debris removal under spatial grasp technology, *Acta Scientific Computer Sciences* 3(7) (2021, July). https://www.actascientific.com/ASCS/pdf/ASCS-03-0135.pdf
105. P.S. Sapaty, Space debris removal under spatial grasp technology, *Network and Communication Technologies* 6(1) (2021). https://www.ccsenet.org/journal/index.php/nct/article/view/0/45486
106. P.S. Sapaty, Spatial grasp model for management of dynamic distributed systems, *Acta Scientific Computer Sciences* 3(9) (2021). https://www.actascientific.com/ASCS/pdf/ASCS-03-0170.pdf
107. P.S. Sapaty, Spatial grasp model for dynamic distributed systems, *Mathematical Machines and Systems* 3 (2021). http://www.immsp.kiev.ua/publications/articles/2021/2021_3/03_21_Sapaty.pdf
108. P.S. Sapaty, Development of space-based distributed systems under spatial grasp technology, *Mathematical Machines and Systems* 4 (2021).

Chapter 2

Satellite constellations, projects, and debris

2.1 INTRODUCTION

The space around Earth is being extensively used by numerous projects from different countries. With the number of satellites, especially in low Earth orbits (LEO), predicted to grow dramatically in the coming years due to the launch of planned satellite constellations, which, despite high attractiveness also raises enormous amount of problems for effective implementation [1–10]. A number of past and currently developing space projects often relate to global defense like Strategic Defense Initiative (SDI) in the past [11–14] with multiple mini-satellites called brilliant pebbles, and the New Space Architecture of the Space Development Agency (SDA) [15–17]. Different commercial projects are being developed too [18–29], including those oriented on global communications, weather prediction, space industry, and robotics. There are also millions of pieces of space junk flying around the Earth, and especially in LEO [10, 30–48]. Their number may be rapidly increasing due to the intensive launch of multi-satellite constellations (sometimes even called "mega-constellations" for their sizes) for very different purposes and particularly in LEO, especially when these satellites come to the end of service or collide with other satellites or the existing junk. This chapter provides a brief review on the growing satellite constellations, existing and expected projects using them, and also rapidly growing space junk which may endanger any further activities in space. The rest of it is organized as follows.

Section 2.2 provides a summary of existing publications on satellite constellations and mega-constellations with their features and problems. These include proper management, using onboard intelligence, providing replacement, issues of effective communications which may be optical, complexity of ground antennas and gateways to monitor low orbit satellites, and also the increase of space debris by constellation units if not de-orbited at the end of service or after failure. *Section 2.3* briefs some defense and industrial projects with multiple satellites in space such as Strategic Defense Initiative of the eighties, and the most recent SDA's Next-Generation Space Architecture. Commercial activity and projects are reviewed too. In *Section*

DOI: 10.1201/9781003230090-2

2.4 a summary on the debris problems and existing solutions is provided with mentioning such issues as legal questions of junk removal, debris surveillance and tracking, the removal complexity, already existing removal contracts and techniques, as well as first removal missions. *Section 2.5* concludes the chapter, stressing the need for a unified system approach for dealing with different types of satellites constellations, especially those on LEO orbits. It also provides references to the Spatial Grasp Technology which will be used throughout the rest of the book for solving different satellite constellations and debris removal problems.

2.2 CONSTELLATIONS AND MEGA-CONSTELLATIONS

We will be giving here a brief review on the rapidly growing numbers of different types of satellites around the Earth and problems of their registration, communication, and management, with more information on that matter easily obtainable in many existing publications, including [1–10].

2.2.1 General on multiple satellites

Near-Earth space is becoming increasingly privatized, with the number of satellites in low Earth orbits predicted to grow dramatically from about 2,000 at present to over 100,000 in the next decade due to the launch of planned satellite constellations [1–4], often even called mega-constellations by their expected size [5–10], as symbolically shown in Figure 2.1.

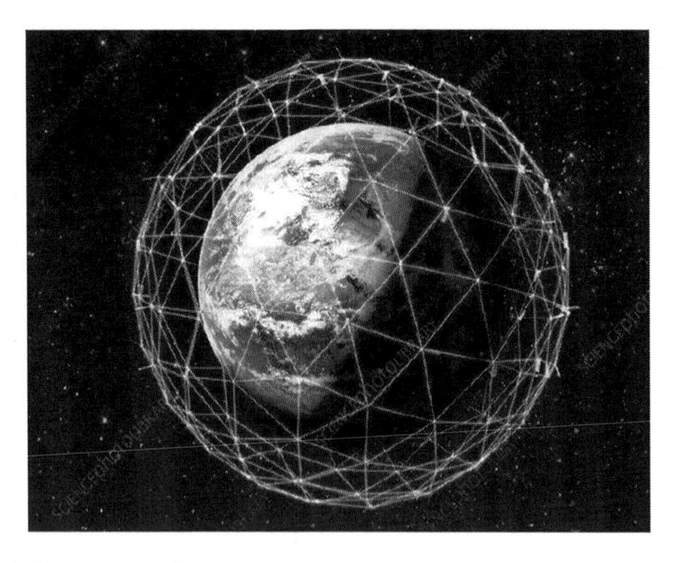

Figure 2.1 Growing satellite constellations.

They will enhance our daily communications, provide us with unending sources of new information, and enable many new applications. Smaller, cheaper satellites, some just the size of a briefcase, can be arranged in different configurations depending on their goal. Spacecraft constellations are appealing especially in three fields: (a) communications, for global coverage, (b) Earth observation, for near real-time measurements, and (c) space observation for continuous monitoring. They offer worldwide communication links competing with terrestrial cellular networks.

LEOs as satellite systems, used in telecommunications, orbit the Earth between 400 and 1,200 miles above the surface. LEO satellites offer advantages over other forms of satellite communication and are used in data communication and email, video conferencing, paging, and more. LEO-based telecommunication systems are less costly than medium Earth orbit (MEO) or high Earth orbit (HEO) options, which also take longer to orbit the Earth. An example of an MEO would be a navigation satellite, designed for monitoring a specific region. Weather and scientific satellites tend to be HEOs, though some are also used in communications..

LEO satellites move at extremely high speeds, due to the pull of Earth's gravity. For example, NASA's Aqua satellite orbits the Earth in approximately 99 min. at 705 km above the surface. According to NASA, we can contrast the speed of the Aqua satellite with that of a weather satellite, about 36,000 km from Earth's surface. Such a weather satellite completes an orbit in about 23 h, 56 min., and 4 s. For another point of reference, consider that the Moon, at 384,403 km from Earth's center, completes an orbit in 28 days.

Data is transmitted through LEO as it is handed off from one satellite to another. LEO satellites typically move in and out of the range of stations transmitting from Earth, which do not need to be as powerful as those transmitting to MEOs or HEOs. This gives LEOs a significant advantage. In addition to requiring less powerful amplifiers for transmission from Earth, LEOs also require less energy to be put into orbit. For this reason, most communication applications use LEO satellites. However, it's important to note that LEO orbits require a network of satellites for continuous coverage.

2.2.2 Constellation management issues

With such a number of active elements in orbit, their management is a fundamental point of interest. Since constellations have not been widely used in the past, not too much work has been done in this sense. Proposed solutions fall into two main categories: (a) optimization of automatic satellite tracking (such as telemetry download), and (b) automatic failure detection, so that the operator does not need to manually check the satellite's status of health. Of much help would be increasing automation onboard the spacecraft, which is not always easy due to satellite size constraint, and prediction, planning, diagnosis, repair, etc., may be deployed on ground. The need for automation is also linked to the collision avoidance assessment and maneuver planning, which are now largely manual. Mostly enabled by technology miniaturization, satellite constellations require a coordinated effort to face the technological limits in spacecraft operations and space traffic. So far, no cost-effective infrastructure is available to withstand coordinated flight of large fleets of satellites.

2.2.3 Onboard intelligence needed

A number of proposals relate an onboard automated managements system based on artificial intelligence. In their implementation, failures may not only be detected, but should also be handled automatically toward a resolution along with a rescheduling of the original plan. The main drawback of such proposals is the need for intersatellite links, which are usually not affordable for low-cost strategies. Moreover, it implies intensive intersatellite communication contributing to RF spectrum crowding. The automatic replanning decreases the ground workload, allowing the operator to concentrate more on the goal, rather than the path to it.

2.2.4 Replacement and variable size of satellites

The replacing strategy is an operational aspect that must be taken into account from the beginning: a replacing or spare strategy is the policy adopted by the operators to substitute failed or terminated satellites of the constellation. These spare satellites are then moved into the constellation when needed. It makes the constellation size adaptable to the market reaction, which is very difficult to predict to configure the orbits in space avoiding multiple launches.

2.2.5 Communication issues

Onboard automation is so far difficult to grow to the point of allowing fully autonomous fleet management, as the large amount of satellites will need to communicate frequently with ground. Overcrowded RF spectrum may cause physical interference of adjacent RF signals. Sharing and integration between space and ground communication terrestrial networks will be effective. The efficient implementation of intersatellite link should be done through routing algorithms taking into account maximum available link time and remnant bandwidth to increase the total traffic capacity of the network in the presence of handover. A completely different approach to avoid RF spectrum overcrowding consists of moving to the optical part of the spectrum. Optical communication promises higher data rates using smaller and lighter terminals, even though due to its high sensitivity to atmospheric conditions it is more suited for free-space intersatellite links rather than satellite-to-ground ones.

2.2.6 Gateways and antennas

High-duty, rugged, and smart Earth terminals and gateways will be needed. Because of the volume of LEO satellites that will soon be flying, the speed at which they're traveling, and variations in frequencies, tracking LEO satellites is challenging. Any terminal or gateway communicating with a LEO satellite will need to receive satellite positions on a regular basis, and this information is pushed to terminals continually. All new LEO constellations will require gateways for tracking the antennas, downloading data, and sending information back to each satellite. Depending on the frequency, gateway antennas vary in size and complexity. The higher the frequency, the harder it is to position the antenna to track and communicate with each satellite. And the larger the constellation, the more terminals or gateways will be needed to maintain frequent communications with each satellite. However, delivering real-time interactive broadband communication services with a large number of LEO satellites, traveling at high speeds over the horizon, requires significantly more complex networks and user terminals than GEO systems. For example, the antennas need to track the moving satellites, and the system needs to handle handover of communication sessions between satellites, whereas just one GEO satellite is required to serve its respective purpose. Constellations with intersatellite links will require fewer gateways. These constellations will be able to maintain private satellite-to-satellite links and minimize the need for continual communication from each satellite to a gateway.

2.2.7 Mega-constellations and mega-debris

Debris experts are worried about an increase in collisions as the number of spacecraft increases, creating even more debris, with a symbolic debris picture shown in Figure 2.2.

The concern for mega-constellations is not about the spacecraft themselves, but rather it is the potential for debris generation from the explosion of or collisions involving the mega-constellation spacecraft [9, 10]. In any case, both debris experts and mega-constellation developers are aware of the additional risks that hundreds or thousands of new satellites could pose for a region of space where the amount of debris continues to grow even with the current population of satellites. Debris is already a problem that is being faced actively with surveillance networks and avoidance maneuvers by the spacecraft operators. An Earth observation satellite may find unexpectedly another one in its field of view, or a region of space may become so overcrowded as to impact the quality of space observations from ground. Making the spacecraft reenter at the end of its life and thus not become debris itself is also of great importance. More on the debris will be discussed in Section 4, which is particularly oriented on debris and problems of their registering, preventing, and removal.

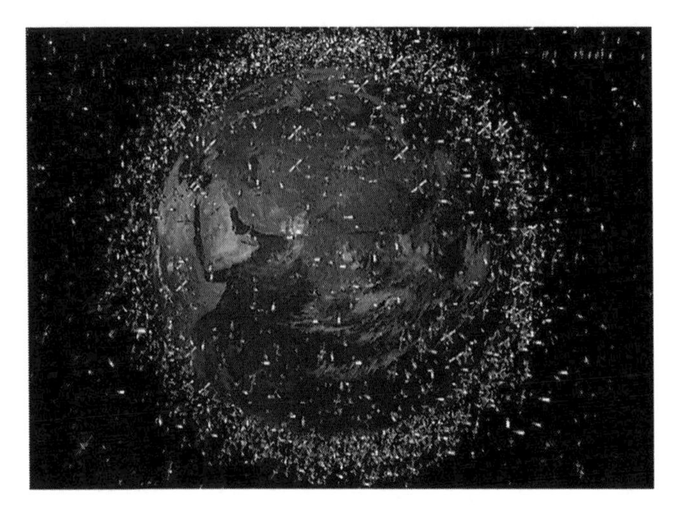

Figure 2.2 Space debris.

2.3 EXAMPLES OF PROJECTS WITH MULTIPLE SATELLITES IN SPACE

This section provides some examples of the current and planned use of multiple satellites, especially in low Earth orbits, with many more obtainable from numerous existing publications, including [11–29].

2.3.1 Strategic defense initiative

The Strategic Defense Initiative (SDI) was a long-term technology research program developed to examine the feasibility of developing defenses against a ballistic missile attack [11–14]. The SDI program was officially launched in 1984, with its key component called Brilliant Pebbles [13, 14], as a proposed space-based weapon for the Global Protection against Limited Strikes (GPALS) Strategic Defense System, see Figure 2.3.

It entailed hundreds of individual interceptors in orbit around the Earth at relatively evenly spaced intervals. Each interceptor could be linked by communications to the others and to ground stations. In the event of a ballistic missile attack, each could be given a high degree of autonomy to detect and intercept missiles that enter its battle space. A set of deployed Brilliant Pebbles would be made up of several staggered rings orbiting at about 400 km above the Earth, with several Brilliant Pebbles in each ring. Once enabled by human command, Brilliant Pebbles could select their targets and divert from their orbits into the path of enemy missiles. The interceptors would carry no explosives, but the force of their high-speed collision is expected to destroy targets.

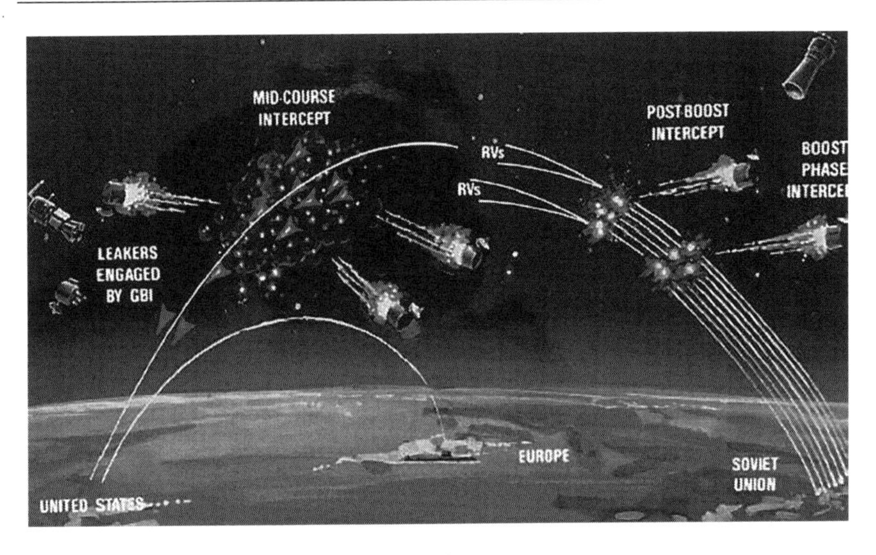

Figure 2.3 Strategic defense initiative architecture.

Brilliant Pebbles were actually thousands of tiny satellites orbiting the Earth. Each was actually a little rocket, no more than 40 inches long, and capable of tracing objects in space on a minutes warning. If an enemy launches nuclear ballistic missiles, and the attack is detected, the tiny space rockets are activated and they race toward the enemy missiles. Most Brilliant Pebbles would intercept and destroy enemy missiles in the so-called "boost" and "post-boost" phases of flight, which occur for up to 5 min. after launch. With additional sensor capabilities the interceptors even could destroy missile warheads during the "mid-course" stage of fight, between 5 and 25 min. after launch, when warheads were moving over great distances and at high speed through space. Mid-course intercept, however, was not a requirement for Brilliant Pebbles at that time. Brilliant Pebbles sensors would allow each interceptor to operate autonomously. As it was able to detect and track enemy missiles on its own, it did not require that outside information, collected by other sensors, to be fed to it.

Brilliant Pebbles interceptors also relied on powerful rocket motors to propel the missile, and steer it into the path of an enemy missile. The interceptor was designed to speed at some 4 mi per second in space in the closing moments of the chase just before impact. Small rockets on the sides of the interceptors called "lateral thrusters" could fire bursts as brief as hundreds of milliseconds, enabling Brilliant Pebbles to maneuver with high degree of precision. What made Brilliant Pebbles particularly safe is that they carried no explosive warhead. They could destroy enemy missiles by the sheer force of collision with them.

Protecting each Brilliant Pebble from the flying debris, dust, and extreme temperatures of space would be what designers call a 'life jacket." It would consist of a solar energy collector panel and rechargeable battery to provide electrical power for the interceptor. The life jacket would be shed by the interceptor after it is ordered into battle, right before it is fired at the incoming ballistic missile. Most important, Brilliant Pebbles had to destroy enemy ballistic missiles when they were most vulnerable, as in the first stages in their flight, and are still carrying their many warheads, which can only be released later in the flight.

2.3.2 Next-generation space architecture

The Next-Generation Space Architecture was recently launched by SDA notional architecture [15–17] (see Figure 2.4). This space architecture plans to fight growing space-based threats, move quickly on hypersonic defense and track hypersonic threats from space, also arm satellites with lasers to shoot down missiles, and so on. Unlike the SDI project, this architecture is oriented on intensive cooperation and collective behavior of many satellites, thus appearing to be of much higher interest for the application of SGT model and technology described in the previous section. The architecture is made up of several layers, as follows.

A *space transport layer* is a global mesh network providing data and communications 24/7. SDA's Transport Layer will provide assured, resilient, low-latency military data and connectivity worldwide to the full range of warfighter platforms. This layer is envisioned, modeled, and architected as a constellation varying in size from 300 to more than 500 satellites in LEO ranging from 750 km to 1200 km in altitude. With a full

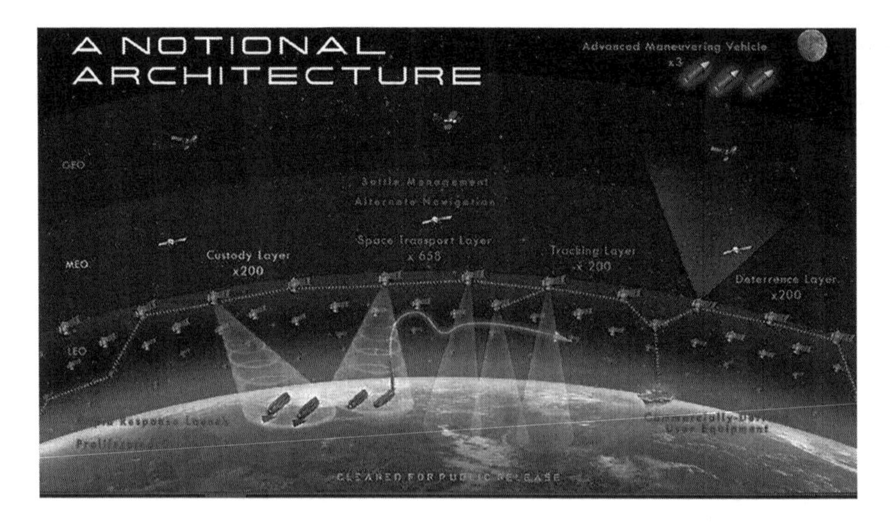

Figure 2.4 Next-generation space architecture.

constellation, 95% of the locations on the Earth will have at least two satellites in view at any given time while 99% of the locations on the Earth will have at least one satellite in view.

A ***tracking layer*** provides tracking, targeting, and advanced warning of missile threats. SDA's tracking layer will provide global indications, warning, tracking, and targeting of advanced missile threats, including hypersonic missile systems. This capability encompasses space-based sensing, as well as algorithms, novel processing schemes, data fusion across sensors and orbital regimes, and tactical data products able to be delivered to the appropriate user.

A ***custody layer*** provides "all-weather custody of all identified timecritical targets." SDA's custody layer will provide 24/7, all-weather custody of time-sensitive, left-of-launch surface mobile targets to support targeting for advanced weapons. The SDA custody layer will maintain track for left-of-launch time-critical targets and enable the creation of a targeting solution from the sensor to the warfighter in operationally relevant timelines through increasing levels of automation.

A ***navigation layer*** provides alternative positioning, navigation, and timing services in case GPS is blocked or unavailable. This layer will provide alternate positioning, navigation, and timing (PNT) for potential Global Positioning System (GPS)-denied environments. The SDA navigation layer will allow for alternate Positioning, Navigation, and Timing (PNT) capabilities for GPS-denied environments supporting both determining predicted orbit and clock offsets, and provisions for transmitting a navigation waveform to DoD and other tactical users.

A ***battle management layer*** is a command, control, and communication network augmented by artificial intelligence that provides self-tasking, self-prioritization, on-board processing, and dissemination. This layer will provide automated space-based battle management through command and control, tasking, mission processing and dissemination to support time-sensitive kill chain closure at campaign scales. It will provide the hardware and software framework to host mission-specific processing, algorithms, and applications across the architecture layers and systems, and will also include space-based command and control, tasking, and mission processing.

A ***support layer*** includes ground command, control facilities, and user terminals, as well as rapid-response launch services. This layer will enable ground systems and launch capabilities to support a responsive and resilient space architecture. The SDA support layer will enable a common, resilient ground support infrastructure necessary to enable the space-based capabilities of the other layers to transmit, receive, process, exploit, and disseminate data. This infrastructure must manage the constellation, host and execute tactical command and control (C2) capabilities, host mission processing and applications.

A ***deterrence layer*** will incubate new mission concepts, its focus will be to deter hostile action in deep space (beyond Geosynchronous Earth Orbit (GEO)) up to lunar distances, and create foundation for new mission concepts that will provide warfighter benefits. The initial focus is to deter an adversary's hostile action in the increasingly active region extending beyond the geosynchronous belt to lunar ranges

2.3.3 Commercial and industrial projects

Mentioning here only some of numerous and rapidly growing activities in this area, with many more available from the existing publications, including [18–29]. NASA plans for commercial LEO development to achieve a robust low Earth orbit economy from which it can purchase services as one of many customers [18]. NASA is opening the International Space Station for commercial business, unleashing U.S. industry on the path to a commercial economy in low Earth orbit. Commercial companies will play an essential role in establishing a sustainable presence in low Earth orbit as well as on and around the Moon, working with NASA to test technologies, train astronauts, and strengthen the burgeoning space economy. Massive commercial space push and a variety of new robotic capabilities will contribute to the rapidly growing space economy and industry [19–22, 29].

Several companies will collectively be launching about 20,000 satellites over the next few years. SpaceX, OneWeb, Telesat, O3b Networks, and Theia Holdings—all have plans to field constellations of V-band satellites in nongeosynchronous orbits to provide communications services in the United States and elsewhere. Space remains a critical frontier for strategic and economic competition among states, and this effort continues to produce significant technological advances [20]. The list of states with national space programs has increased markedly. Europe has pooled its efforts through the European Space Agency and has collectively derived a benefit of scale through EU programs such as Galileo and Copernicus. China, India, and Japan are now prominent players, and a number of other countries have also entered the field. International cooperative programs have resulted in new links being created, often in the form of common projects.

LEO economy encompasses Earth-centered orbits with an altitude of 2,000 km (1,200 mi) or less [21, 22], and is considered for convenient transportation, communication, observation, and resupply. This is the area where the International Space Station currently orbits and where many proposed future platforms will be located with engagement in production, distribution, and trade of goods and services. As the technology progresses, this economic space will grow with more governmental, commercial, academic, and other groups that will contribute to the LEO economy's continued expansion and support of future sustainable space enterprises.

There has been resurgent interest in building large LEO constellations of satellites on a new and bigger scale, including communication [23, 24]. The aim is delivering internet to the world by providing truly global and robust broadband coverage. Players such as OneWeb, SpaceX, Samsung, and Boeing all have proposals for such a system. Each plans on launching constellations of 600 to over 4,000 satellites, dwarfing the 1,400 operational satellites currently in orbit. This sheer number of satellites, along with their global coverage, gives rise to opportunities not only for broadband but also as a platform for providing navigation services.

2.4 DEBRIS PROBLEMS AND SOLUTIONS

We will be giving here a brief review on debris and problems of dealing with them, with more information available in the existing publications in this area, including [30–48].

2.4.1 General on debris

The threats from the existing numerous space debris are becoming enormously high [30–32]. The trash situation caused by intensive new launches and defunct satellites is getting worse and risks making space off-limits for future generations [31] (see Figure 2.5 for some symbolic space junk pictures). Space junk is no one country's responsibility [32], but the responsibility of every space-faring country.

The problem of managing space debris is both an international challenge and opportunity to preserve the space environment for future space exploration missions. LEO is an orbital space junk yard, with millions of pieces of space junk flying there. Most orbital debris comprises human-generated objects, such as pieces of space craft, tiny flecks of paint from a spacecraft, parts of rockets, satellites that are no longer working, or explosions of objects in orbit flying around in space at high speeds. The threats of space debris are increasing due to the launch of several multi-satellite constellations, particularly in LEO. The new space paradigm and the increasing

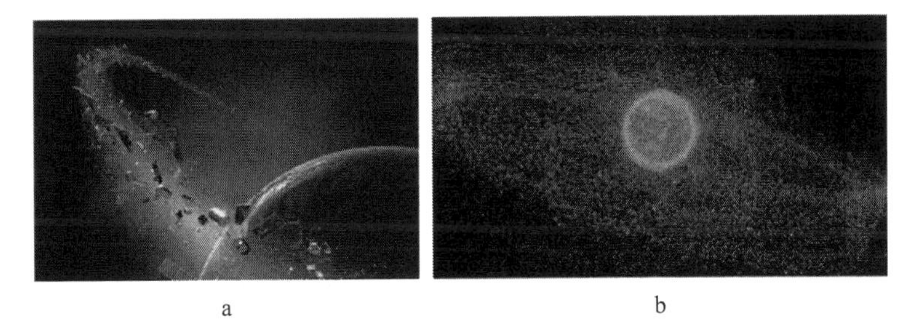

a b

Figure 2.5 Space debris: (a) certain orbits related, (b) overall picture.

population of spacecraft in LEO require de-orbiting systems that can satisfy space debris requirements [33]. Drag sails are the main technology, and several companies have already commercialized and sold these products. Other systems such as electromagnetic tethers, deployable booms, or the NASA Exo-brake have also already been prototyped and demonstrated in space.

2.4.2 Legal issues of removal

Various legal and political concepts to resolve the problem of the existing space debris in outer space are analyzed, also which measures to take to avoid space debris or to reduce potential space debris in the course of future space missions [34]. From a scientific and technical point of view various studies are ongoing to analyze the feasibility of active debris removal. Nevertheless it has to be highlighted that outer space is an international area where various actors with different legal and political concepts are operating, a situation that leads to different approaches concerning such activities. Space debris is the global mounting ultimatum to the enduring maintainability of the outer space activities, and it ought to be dealt in the very beginning, otherwise, it will be too late [35]. From couple of years ago, some incidents of collisions have enhanced the space debris accumulation, now crowded the corridor of Earth orbit which constitutes the most serious pollutant of the near-Earth space environment.

2.4.3 Surveillance and tracking

Currently, over 22,000 objects larger than 10 cm are tracked by Space Surveillance Networks and recorded in their catalog to provide warnings to satellites in the path of these objects and to enable them to perform avoidance maneuvers [36]. From a technical point of view, here the challenges are the identification, tracking, and cataloging of the centimeter-sized objects, still large enough to produce catastrophic damage but not included in the current catalogs. Radars have been the preferred ground-based system, in particular to monitor LEO, as they can operate independently day and night as well as in all meteorological conditions. However, most radar telescopes are optimized for astronomical observations rather than debris tracking and so bistatic systems have also been used to improve performance, and some have shown a capability to detect objects down to 1 cm at 100 km. Similarly, systems combining laser ranging and passive optical tracking have been demonstrated to achieve good accuracy in determining the position of objects (within 10 m). These capabilities and different organization catalogs have to come together to improve actionable knowledge of the orbital population.

2.4.4 Complexity of removal

Despite promising technology demonstrations, there is no one-size-fits-all solution for the growing problem of taking out the orbital trash [37]. Even tiny pieces of space debris can have catastrophic effects. A Space Age "tragedy of the commons" is unfolding right under our nose (or, really, right over our head) and no consensus yet exists on how to stop it. For more than half a century, humans have been hurling objects into LEO in ever growing numbers. And with few meaningful limitations on further launches into that increasingly congested realm, the prevailing attitude has been persistently permissive: in orbit, it seems, there is always room for one more.

2.4.5 Removal contracts and techniques

NASA and ESA studies show that the only way to stabilize the orbital environment is to actively remove large debris items [38–40]. ClearSpace-1 will be the first space mission to remove an item of debris from orbit, planned for launch in 2025, as the first-ever space mission to clean orbital junk with the use of a giant claw. The first-of-its-kind mission is not only new in terms of what it's setting out to achieve, but also represents a shift in strategy for the ESA, which has chosen a private firm to design and engineer its own spacecraft and plan of execution. A Japanese company said recently it will develop a satellite to clean up floating space debris by using laser beams, with the aim of starting the service in 2026 [41]. Satellite communications company Sky Perfect JSAT Corp. said the project will be the first to use laser beams to remove space debris such as defunct satellites and rocket sections. To preserve a secure space environment, the active removal or de-orbiting of space debris is an emergent technological challenge. If remedial action is not taken in the near future, it will be difficult to sustain human space activities. To overcome this issue, several other methods for the removal and de-orbiting of debris have been proposed so far; classified as either contact (e.g., robotic arm, tether net, electrodynamic tether) or contactless ones like plasma beam [42] ejected from the satellite to impart a force to the debris thereby decelerating it, which results in it falling to a lower altitude, reentering the Earth's atmosphere and burning up naturally (see Figure 2.6 for some existing junk-cleaning techniques).

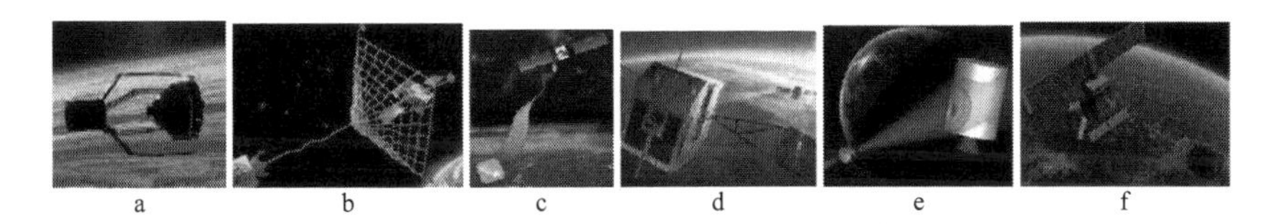

Figure 2.6 Active deorbiting techniques: (a) mechanical grasp, (b) net, (c) tether, (d) laser, (e) ion beam, (f) directed energy.

2.4.6 Very first removal missions

The spacecraft and the 17-kilogram dummy satellite (the debris to be cleaned up) will separate and then perform a high-stakes game of cat and mouse [43–45]. A demonstration mission to test new technology developed by the company Astroscale to clean up space debris is to be launched from the Baikonur Cosmodrome in Kazakhstan. Known as ELSA-d, the mission will exhibit technology that could help capture space junk, the millions of pieces of orbital debris that float above Earth. The more than 8,000 metric tons of debris threaten the loss of services we rely on for Earth-bound life, including weather forecasting, telecommunications, and GPS systems. Some research on methods of debris analysis and removal can be found in [46–48].

2.5 CONCLUSIONS

Concluding this review chapter, we are stressing here the need for a unified system approach for dealing with different types of satellites constellations, especially those in LEO orbits. In the following chapters the use of the developed Spatial Grasp Technology (SGT) [49–72] for dealing with satellite constellations and different projects based on them will be discussed in detail. Of special interest are large constellations of relatively simple and cheap LEO satellites which need special management models and approaches due to rapidly changing their positions over Earth locations. To provide continuous Earth observation, they must be in large numbers, work cooperatively, and regularly transfer their duties and accumulated information to other satellites. We must also consider different possibilities for organization of massive removal of most dangerous junk items, especially in the LEO orbits, and by large constellations of cleaning satellites, as the only working solution. Of course, such mega-cleaning constellations do not exist yet, and unfortunately many private companies and governmental organizations are chaotically launching numerous and cheap satellites (which soon will count up to 100,000 in LEO). But the danger of uncontrolled rubbishing of space around Earth may be very high and lead to even more severe consequences than the global warming and current worldwide COVID disaster.

REFERENCES

1. United Nations Register of Objects Launched into Outer Space, *The United Nations Office for Outer Space Affairs*. http://www.unoosa.org/oosa/en/spaceobjectregister/index.html
2. G. Curzi, D. Modenini, P. Tortora, Review Large Constellations of Small Satellites: A Survey of Near Future Challenges and Missions, *Aerospace* 7(9) (2020), 133. https://www.mdpi.com/2226-4310/7/9/133/htm
3. K. Dredge, M. von Arx, I. Timmins, LEO constellations and tracking challenges. www.satellite-evolution.com, September/October 2017.

4. A. Venkatesan, J. Lowenthal, P. Prem, M. Vidaurri, The impact of satellite constellations on space as an ancestral global commons, *Nature Astronomy* 4 (2020, November), 1043–1048. www.nature.com/natureastronomy
5. R. Skibba, How satellite mega-constellations will change the way we use space, *MIT Technology Review*, February 26, 2020. https://www.technologyreview.com/2020/02/26/905733/satellite-mega-constellations-change-the-way-we-use-space-moon-mars/
6. M. Minet, The space legal issues with mega-constellations, November 3, 2020. https://www.spacelegalissues.com/mega-constellations-a-gordian-knot/
7. E. Siegel, Astronomy faces a mega-crisis as satellite mega-constellations loom, January 19, 2021. https://www.forbes.com/sites/startswithabang/2021/01/19/astronomy-faces-a-mega-crisis-as-satellite-mega-constellations-loom/?sh=30597dca300d
8. A. Jones, China is developing plans for a 13,000-satellite megaconstellation, *Space News*, April 21, 2021. https://spacenews.com/china-is-developing-plans-for-a-13000-satellite-communications-megaconstellation/
9. N. Reilanda, A. J. Rosengren, R. Malhotra, C. Bombardelli, Assessing and minimizing collisions in satellite mega-constellations, 2021, Published by Elsevier B.V. on behalf of COSPAR. https://www.sciencedirect.com/science/article/abs/pii/S0273117721000326
10. J. Fous, Mega-constellations and mega-debris, October 10, 2016. https://www.thespacereview.com/article/3078/1
11. Strategic Defense Initiative. *The White House*. 1984. https://fas.org/irp/offdocs/nsdd/nsdd-119.pdf
12. The Strategic Defense Initiative: Program Facts. 1987. https://www.everycrsreport.com/files/19870722_IB85170_64d13e614c37eecbed39c00741ddfb269f814fef.pdf
13. J. Gattuso, Brilliant Pebbles: The Revolutionary Idea for Strategic Defense. 1990. https://www.heritage.org/defense/report/brilliant-pebbles-the-revolutionary-idea-strategic-defense
14. Brilliant Pebbles. https://en.wikipedia.org/wiki/Brilliant_Pebbles
15. Space Development Agency Next-Generation Space Architecture. 2019. https://www.airforcemag.com/PDF/DocumentFile/Documents/2019/SDA_Next_Generation_Space_Architecture_RFI%20(1).pdf
16. Magnuson S. Web Exclusive: Details of the Pentagon's New Space Architecture Revealed. 2019. https://www.nationaldefensemagazine.org/articles/2019/9/19/details-of-the-pentagon-new-space-architecture-revealed
17. Messier D. Space Development Agency Seeks Next-Gen Architecture in First RFI. 2019. http://www.parabolicarc.com/2019/07/07/space-development-agency-issues-rfi/
18. NASA Plan for Commercial LEO Development to achieve a robust low-Earth orbit economy from which NASA can purchase services as one of many customers, Summary and Near-Term Implementation Plans, June 7, 2019. https://www.nasa.gov/sites/default/files/atoms/files/commleodevt_plan_6-7-19_final-links-new.pdf
19. Massive commercial space push and a variety of new robotic capabilities could self supporting and rapidly growing space economy. https://www.nextbigfuture.com/2017/03/massive-commercial-space-push-and.html
20. J.-M. Bockel, The Future of the Space Industry. *General Report*, 2018. November 17. https://www.nato-pa.int/download-file?filename=sites/default/

files/2018-12/2018%20-%20THE%20FUTURE%20OF%20SPACE%20INDUSTRY%20-%20BOCKEL%20REPORT%20-%20173%20ESC%2018%20E%20fin.pdf
21. G. Martin, NewSpace: The «Emerging» Commercial Space Industry. https://ntrs.nasa.gov/archive/nasa/casi.ntrs.nasa.gov/20140011156.pdf
22. LEO Economy FAQs. https://www.nasa.gov/leo-economy/faqs
23. NASA Wants Orbital Facility Designs for Commercial LEO Marketplace Project. https://www.govconwire.com/2021/07/nasa-wants-orbital-facility-designs-for-commercial-leo-marketplace-project/
24. Leveraging Commercial Broadband LEO Constellations for Navigating. https://www.ion.org/publications/abstract.cfm?articleID=14729
25. Proliferated Commercial Satellite Constellations Implications for National Security. https://ndupress.ndu.edu/Portals/68/Documents/jfq/jfq-97/jfq-97_20-29_Hallex-Cottom.pdf?ver=2020-03-31-130614-940
26. How low earth orbit satellites are changing the game. https://x2n.com/blog/how-low-earth-orbit-satellites-are-changing-the-game/
27. Aboard Commercial Rocket, Space Defense Agency Sends Up Satellites for First Time. https://www.sda.mil/aboard-commercial-rocket-space-defense-agency-sends-up-satellites-for-first-time/
28. #SpaceWatchGL Opinion: Orbofleet's Take on Mega-constellations and LEO Satellite Market. https://spacewatch.global/2021/07/orbofleets-take-on-mega-constellations-and-leo-satellite-market/
29. Min Read Space robotics market to reach $3.5bn by 2025: GMI report. 2019. May 12. https://satelliteprome.com/news/space-robotics-market-to-reach-3-5bn-by-2025-reveals-gmi-report/
30. Space Debris. https://en.wikipedia.org/wiki/Space_debris
31. P. Teffer, Europe's space trash chief: situation getting worse, Interview, EUOBSERVER, August 2018. https://euobserver.com/science/142685
32. Space Debris, NASA Headquarters Library. https://www.nasa.gov/centers/hq/library/find/bibliographies/space_debris
33. Deorbit Systems, National Aeronautics and Space Administration, November 28, 2020. https://www.nasa.gov/smallsat-institute/sst-soa-2020/passive-deorbit-systems
34. A. Froehlich (Ed.), *Space Security and Legal Aspects of Active Debris Removal*, Springer, 2019. https://www.springer.com/gp/book/9783319903378
35. A. Sheer, S. Li, Space debris mounting global menace legal issues pertaining to space debris removal: Ought to revamp existing space law regime, *Beijing Law Review* 10 (2019), 423–440. https://www.scirp.org/pdf/BLR_2019051615104007.pdf
36. G. S. Aglietti, From space debris to NEO, some of the major challenges for the space sector, *Frontiers Space and Technology*, June 16, 2020. https://www.frontiersin.org/articles/10.3389/frspt.2020.00002/full
37. L. David, Space junk removal is not going smoothly, *Scientific American*, April 14, 2021. https://www.scientificamerican.com/article/space-junk-removal-is-not-going-smoothly/
38. ESA commissions world's first space debris removal, *ESA / Safety & Security / Clean Space*, December 9, 2019. https://www.esa.int/Safety_Security/Clean_Space/ESA_commissions_world_s_first_space_debris_removal
39. A. Parsonson, ESA signs contract for first space debris removal mission, *Space News*, December 2, 2020. https://spacenews.com/clearspace-contract-signed/

40. Humza, The first-ever space mission to clean orbital junk will use a giant claw, *Techspot*, December 1, 2020. https://www.techspot.com/community/topics/the-first-ever-space-mission-to-clean-orbital-junk-will-use-a-giant-claw.266509/
41. Japanese company planning space debris removal by laser on satellite, KYODO NEWS, August 8, 2020. https://english.kyodonews.net/news/2020/08/fc06829d1d9a-japanese-company-planning-space-debris-removal-by-laser-on-satellite.html
42. Plasma thruster: New space debris removal technology, Tohoku University, 27 September, 2018. https://www.eurekalert.org/pub_releases/2018-09/tu-ptn092718.php
43. K. Hunt, Mission to clean up space junk with magnets set for launch, CNN, 1 April, 2021. https://edition.cnn.com/2021/03/19/business/space-junk-mission-astroscale-scn/index.html
44. M. Obe, Japan's Astroscale launches space debris-removal satellite, *Nikkei Asia*, 22 March, 2021. https://asia.nikkei.com/Business/Aerospace-Defense/Japan-s-Astroscale-launches-space-debris-removal-satellite
45. C. Weiner, New effort to clean up space junk reaches orbit, March 21, 2021. https://www.npr.org/2021/03/21/979815691/new-effort-to-clean-up-space-junk-prepares-to-launch
46. Y. Chen et al., Optimal mission planning of active space debris removal based on genetic algorithm, IOP Conf. Series: Materials Science and Engineering 715 (2020), 012025. https://iopscience.iop.org/article/10.1088/1757-899X/715/1/012025/pdf
47. R. Klima et al., Space debris removal: Learning to cooperate and the price of anarchy, *Frontiers in Robotics AI*, 04 June, 2018. https://www.frontiersin.org/articles/10.3389/frobt.2018.00054/full
48. Research on space debris, safety of space objects with nuclear power sources on board and problems relating to their collision with space debris. Committee on the Peaceful Uses of Outer Space. Vienna, 2019. http://www.unoosa.org/res/oosadoc/data/documents/2019/aac_105c_12019crp/aac_105c_12019crp_7_0_html/AC105_C1_2019_CRP07E.pdf
49. P.S. Sapaty, A distributed processing system, European Patent N 0389655, Publ. 10.11.93, European Patent Office. 35 p.
50. P.S. Sapaty, *Symbiosis of Real and Simulated Worlds under Spatial Grasp Technology*. Springer, 2021. 305 p.
51. P.S. Sapaty, *Complexity in International Security: A Holistic Spatial Approach*. Emerald Publishing, 2019. 160 p.
52. P.S. Sapaty, *Holistic Analysis and Management of Distributed Social Systems*. Springer, 2018. 234 p.
53. P.S. Sapaty, *Managing Distributed Dynamic Systems with Spatial Grasp Technology*. Springer, 2017. 284 p.
54. P.S. Sapaty, *Ruling Distributed Dynamic Worlds*. New York: John Wiley & Sons, 2005. 255 p.
55. P.S. Sapaty, *Mobile Processing in Distributed and Open Environments*. New York: John Wiley & Sons, 1999. 410 p.
56. P.S. Sapaty, Advanced terrestrial and celestial missions under spatial grasp technology, *Aeronautics and Aerospace Open Access Journal* 4(3) (2020). https://medcraveonline.com/AAOAJ/AAOAJ-04-00110.pdf

57. P.S. Sapaty, Spatial Management of Distributed Social Systems, *Journal of Computer Science Research* 2(3) (2020, July). https://ojs.bilpublishing.com/index.php/jcsr/article/view/2077/pdf
58. P.S. Sapaty, Towards Global Nanosystems Under High-level Networking Technology, *Acta Scientific Computer Sciences* 2(8) (2020). https://www.actascientific.com/ASCS/pdf/ASCS-02-0051.pdf
59. P.S. Sapaty, Symbiosis of Distributed Simulation and Control under Spatial Grasp Technology, SSRG *International Journal of Mobile Computing and Application (IJMCA)* 7(2) (2020, May–August). http://www.internationaljournalssrg.org/IJMCA/2020/Volume7-Issue2/IJMCA-V7I2P101.pdf
60. P.S. Sapaty, Global Network Management under Spatial Grasp Paradigm, *International Robotics & Automation Journal* 6(3) (2020). https://medcraveonline.com/IRATJ/IRATJ-06-00212.pdf
61. P.S. Sapaty, Global Network Management under Spatial Grasp Paradigm, *Global Journal of Researches in Engineering: J General Engineering* 20(5) (2020) Version 1.0. https://globaljournals.org/GJRE_Volume20/6-Global-Network-Management.pdf
62. P.S. Sapaty, Symbiosis of Real and Simulated Worlds Under Global Awareness and Consciousness, *The Science of Consciousness Symposium TSC* 2020. https://eagle.sbs.arizona.edu/sc/report_poster_detail.php?abs=3696
63. P.S. Sapaty, Fighting global viruses under spatial grasp technology, *Transactions on Engineering and Computer Science* 1(2) (2020). https://gnoscience.com/uploads/journals/articles/118001716716.pdf
64. P.S. Sapaty, Symbiosis of Virtual and Physical Worlds under Spatial Grasp Technology, *Journal of Computer Science & Systems Biology* 13(6) (2020). https://www.hilarispublisher.com/open-access/symbiosis-of-virtual-and-physical-worlds-under-spatial-grasp-technology.pdf
65. P.S. Sapaty, Spatial grasp as a model for space-based control and management systems, *Mathematical Machines and Systems* 1 (2021), 135–138. http://www.immsp.kiev.ua/publications/articles/2021/2021_1/Sapaty_book_1_2021.pdf
66. P.S. Sapaty, Managing multiple satellite architectures by spatial grasp technology, *Mathematical Machines and Systems* 1 (2021), 3–16. http://www.immsp.kiev.ua/publications/eng/2021_1/
67. P.S. Sapaty, Spatial Management of Large Constellations of Small Satellites, *Mathematical Machines and Systems* 2 (2021). http://www.immsp.kiev.ua/publications/articles/2021/2021_2/02_21_Sapaty.pdf
68. P.S. Sapaty, Global Management of Space Debris Removal under Spatial Grasp Technology, *Acta Scientific Computer Sciences* 3(7) (2021, July). https://www.actascientific.com/ASCS/pdf/ASCS-03-0135.pdf
69. P.S. Sapaty, Space Debris Removal under Spatial Grasp Technology, *Network and Communication Technologies* 6(1) (2021). https://www.ccsenet.org/journal/index.php/nct/article/view/0/45486
70. P.S. Sapaty, Spatial Grasp Model for Management of Dynamic Distributed Systems, *Acta Scientific Computer Sciences* 3(9) (2021). https://www.actascientific.com/ASCS/pdf/ASCS-03-0170.pdf
71. P.S. Sapaty, Spatial Grasp Model for Dynamic Distributed Systems, *Mathematical Machines and Systems* 3 (2021). http://www.immsp.kiev.ua/publications/articles/2021/2021_3/03_21_Sapaty.pdf
72. P.S. Sapaty, Development of Space-based Distributed Systems under Spatial Grasp Technology, *Mathematical Machines and Systems* 4 (2021).

Chapter 3

Spatial Grasp Model (SGM) and Spatial Grasp Technology (SGT)

3.1 INTRODUCTION

This chapter inherits the experience obtained in creation of citywide computer networks in Kiev, Ukraine, from the end of sixties with author's active participation, which were integrating different institutes of the National Academy of Sciences and other organizations, well before the internet. By spreading a fully interpreted scenario code in a wavelike mode between different computers, it was possible to solve complex analytic-numerical problems on heterogeneous computer networks that were difficult to organize on individual computers. These works resulted in a new management concept and real distributed control methodology and technology, which were further developed in different countries with application in such areas as intelligent network management, industry, social systems, psychology, collective robotics, security, and defense. A special high-level recursive Spatial Grasp Language has been developed in which distributed, parallel, and holistic algorithms could be expressed with resultant spatial scenarios up to a hundred times more compact and proportionally simpler than in other languages. All this activity resulted in a European patent and more than 200 international publications, including six books. The aim of the current chapter is to generalize all these works and obtained experience in the form of radically new computational, control, and management model as a natural extension of traditional concept of algorithm and its exhibition by flowcharts, with potential applications in large distributed systems operating in combined terrestrial and celestial environments. This model allows us to express complex solutions in distributed spaces with feeling of direct staying in and moving through them, also to obtain their overall vision and understanding in a holistic manner. This spatial model also allows for the development of radically new distributed and holistic control and management technology to be discussed in this chapter too.

DOI: 10.1201/9781003230090-3

The rest of the chapter is organized as follows. *Section 3.2* briefs the traditional concept of algorithm as a finite sequence of well-defined, computer-implementable instructions, and the widely used notion of flowchart as a type of diagram that represents a workflow or process. *Section 3.3* describes basics of Spatial Grasp model and how it differs from conventional algorithm. This section also introduces a new type of a chart called spatiochart as a further development of traditional flowcharts for describing and analyzing scenarios operating directly in distributed spaces. It shows how collections of actions can be described in SG and exhibited by spatiocharts, including the use of control rules supervising repetition, sequencing and branching in the spatial scenarios, also expressing spatial dataflow and exchange, with such organizations capable of being unlimitedly hierarchical and recursive. *Section 3.4* briefs the main elements of Spatial Grasp Technology (SGT) and its high-level recursive Spatial Grasp Language (SGL) based on SG philosophy, with already existing numerous publications, including books, on this approach, its implementation, and numerous applications. This includes different types of distributed worlds SGT operates with, various constants which may represent both information and physical matter, repertoire of spatial variables of SGL, some of which may be stationary while others mobile, main types of SGL rules which can be nested, different control states provided by SGL scenarios propagation, and general organization of the distributed and networked SGL interpreter. *Section 3.5* concludes the chapter, providing references to the researched, tested, and already published SGT applications in very different areas.

3.2 ALGORITHM AND FLOWCHARTS

Algorithm is a finite sequence of well-defined, computer-implementable instructions, typically to solve a class of specific problems or to perform a computation [1–3]. Algorithms are always unambiguous and are used as specifications for performing calculations, data processing, automated reasoning, and other tasks. In contrast, a *heuristic* is a technique used in problem solving that uses practical methods and/or various estimates in order to produce solutions that may not be optimal but are sufficient given the circumstances [4].

A *flowchart* is a type of diagram that represents a workflow or process [5, 6]. A flowchart can also be defined as a diagrammatic representation of an algorithm, a step-by-step approach to solving a task. The flowchart shows the steps as boxes of various kinds, and their order by connecting the boxes with arrows. Flowcharts are used in analyzing, designing, documenting, or managing a process or program in various fields. Examples of simple flowcharts are shown in Figure 3.1 (where a processing step is usually depicted as a rectangular box and a decision as a diamond).

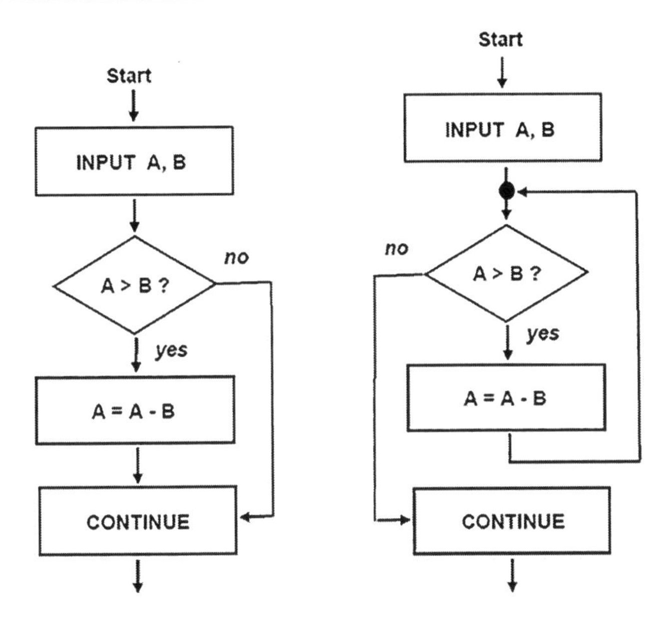

Figure 3.1 Examples of flowcharts.

3.3 SPATIAL GRASP (SG) VERSUS TRADITIONAL ALGORITHM

3.3.1 Elementary Spatial Grasp explanation

SG model [7–13], which may be considered as a *spatial extension and interpretation of the concept of algorithm*, operates by recursive scenarios self-spreading in physical and virtual worlds while creating, matching, transforming, processing and managing them. The interpreted scenario text may not be staying permanently in any point or points (if this not required) and can move in space while carrying further its still unprocessed remainder and omitting utilized parts (if the latter not needed any more). We start explaining SG model on elementary operations, then move to more complex solutions with many actions, and finally come to universal recursive definition of any SG scenarios.

3.3.1.1 Single operation

Imagine you are staying in some point of space (which may be a paper sheet, computer memory, or any place in terrestrial or celestial environment) and just writing:

```
44.55
```

You will get this value in the current world point which may stay there indefinitely, without any name. Another example:

```
5 + 6
```

This operation will produce the value 11 which will stay in this point without a name too. One more:

```
R = 5 + 6
```

The result 11 will be assigned to a variable `R` and will stay in the starting point under this name. It may be subsequently accessed by name `R` if to come into this point again. Other example, but now related to physical space:

```
move(x55_y88)
```

From the current point in space you will move to another point with certain `x_y` coordinates and will stay there. If you want to create a node named `John` in virtual space, with moving to and staying indefinitely in this node, just write:

```
create('John')
```

If the node `John` already exists, you may directly hop into it from the starting point with staying there, as follows.

```
hop('John')
```

A single action may produce a multiple result, for example, by hopping simultaneously to virtual nodes `John`, `Peter`, and `Alex` if they already exist, with staying in all of them, as:

```
hop('John', 'Peter', 'Alex')
```

Or moving in parallel to a number of physical world locations from a starting point with staying in them all indefinitely, as follows:

```
move(x55_y88, x5_y12, x105_y92)
```

Generalizing the above mentioned and other possible examples with a single action, let the latter be named just as `g`, and applied in some `Start` point, we can receive the result in some region of space (symbolically named as `G` too, but in capital), which may include the `Start` position, as in the first examples above. This is shown in Figure 3.2 in the form of a special diagram

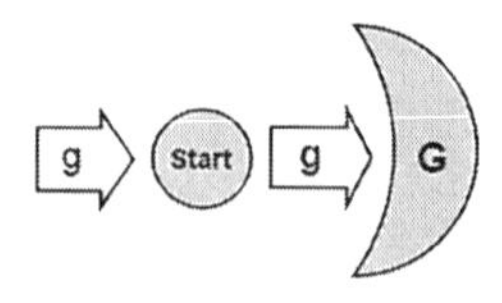

Figure 3.2 The simplest spatiochart.

or chart, which we will be calling *spatio-chart* (or just *spatiochart*), as an extension to traditional *flowchart* but for representing operations and their groupings working directly in distributed spaces, which SG philosophy and model pursue.

3.3.1.2 Sequence of operations

Let us consider now expression of a sequence of possible operations in space, with using semicolon as a delimiter between them.

Assigning to a variable `R` and then changing its value, with finally staying in the same position in space, can be as follows:

```
R = 15; R = R + 10
```

Hopping to virtual node `John` and then creating a new node `Peter` with relation to it from `John` as of Peter's `father`, with final staying in node `Peter`:

```
hop('John'); create(link('father'), node('Peter'))
```

Hopping to virtual node `Peter` and then creating a `friend` relation to the already existing node `John`, with final staying in node `John`:

```
hop('Peter'); linkup('friend', node('John'))
```

Moving to a physical location by its absolute coordinates and then shifting twice to other locations by given coordinate changes, with final staying in the node reached by the second shift:

```
move(x55_y88); shift(x11_y22); shift(x9_y45)
```

The mentioned and any other examples with sequences of actions `gi` initially applied from some `Start` position, with the next action `gi+1` originating in all or some space positions reached by the previous action `gi`, can be represented just as:

```
g1; g2; g3
```

Their combined operation is shown by the spatiochart in Figure 3.3 with regions (symbolically represented by Crescent Moon shapes) reached by

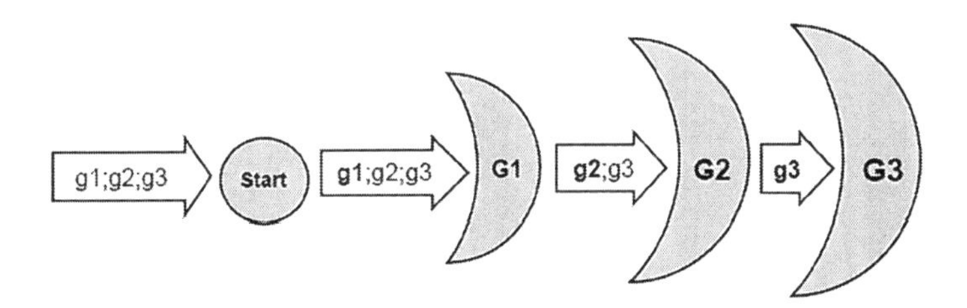

Figure 3.3 Spatiochart for a sequence of actions in space.

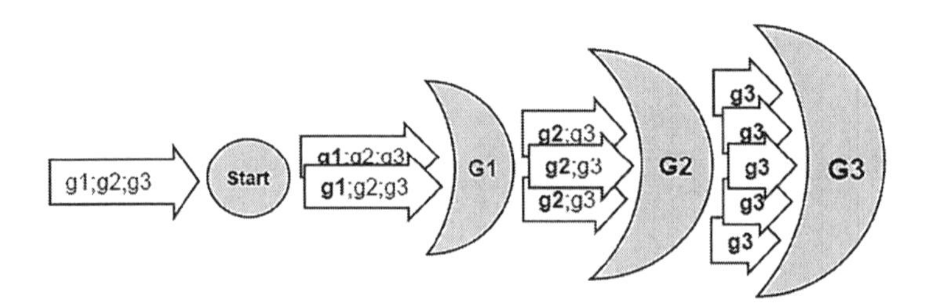

Figure 3.4 Possible code replications during parallel space navigation.

actions `gi` named as `Gi` too (which may generally include positions reached by the previous actions in the previous regions). This also takes into account that the *rest of the sequence has to propagate in space* by delivering descriptions of further actions, and the already used operations *are removed* from the sequence as not needed any more.

In Figure 3.4, which may detail the sequence of operations of Figure 3.3, it is shown that the movement from regions `Gi` to `Gi+1` can be generally made in parallel from different points of `Gi` (from the same points potentially too, as `gi` may themselves represent parallel operations), so the operational sequences in reality can be replicated at any stage of their development.

3.3.2 Using rules

Each operation is usually based on certain *rules* defining its origin, sense, and details of implementation, management, and control. We will consider the concept of rule here in a broad sense, as being itself a certain entity and mechanism which may influence different operations, or even completely represent them.

3.3.2.1 *In sequencing of operations*

In a more advanced organization of the sequence of operations we may use different rules embracing them, which can provide additional (including nonlocal) control, functionality, and more advanced processing and coverage of distributed spaces as, for example, by rule `r1` for the operational sequence considered before:

```
r1(g1; g2; g3)
```

The rule will be activated in the position where the whole sequence is applied, like `Start` as before. It then may influence the whole sequence of embraced operations with receiving a feedback from its entire development (if such feedback is needed by the rule's functionality), as shown in Figure 3.5.

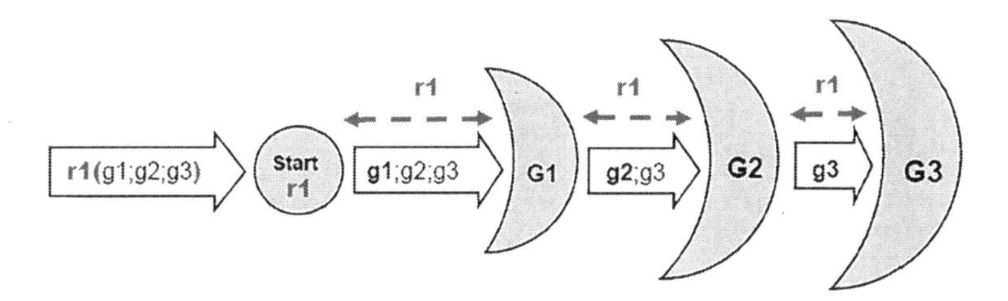

Figure 3.5 Spatiochart with a control rule.

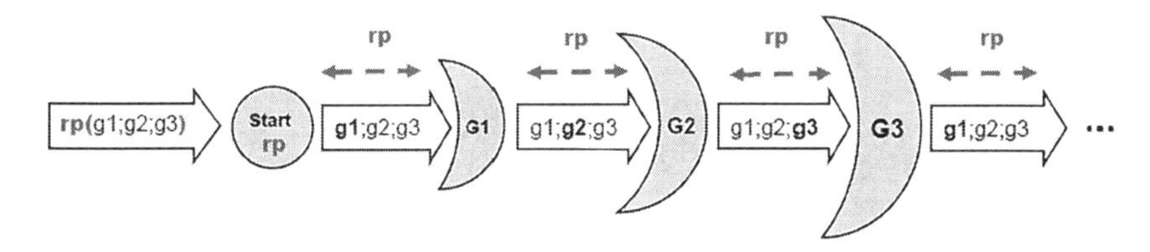

Figure 3.6 Repeated navigation of space.

The rule, for example, may represent such functionality as print, create, repeat, and many other cases of nonlocal influence, management, and control. Let us consider a few examples in more detail.

```
r1 → print
```

By this rule, the results obtained by the embraced sequence of operations (and not only by its last operation `g3` but also by `g1` and `g2`, if producing final results too) can be returned to the `Start` position and printed there, with final staying in the `Start`.

```
r1 → create
```

This rule can supply the space propagating operations, especially those describing movement in virtual spaces, with global power of creating these spaces or their individual elements if they are absent during this movement (and therefore do not allow to proceed further). This means that the same written sequence of actions `gi` can work in both space navigation and space creation modes, depending on circumstances.

```
r1 → repeat
```

Under this rule, the sequence of operations `gi` at first is processed as usual, step by step, until the rest of it becomes empty; but after this, it starts to work from the very beginning again, as shown in Figure 3.6. The `repeat` rule always saves the already processed operations in their sequence (which could be removed without it), with the whole sequence repeatedly propagating and working in space until possible.

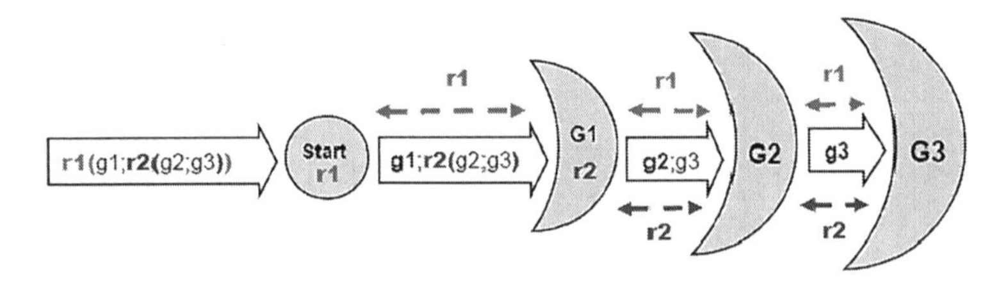

Figure 3.7 Nested control rules.

The operational sequence can be embraced by any number of control rules, which can be nested, as follows for rules `r1` and `r2` and also shown in Figure 3.7 (with `r1` activated in the `Start` position, and `r2` in the positions in space belonging to `G1` (which may be many) and reached by operation `g1`.

```
r1(g1; r2(g2; g3))
```

A few examples of combination of such nested rules:

```
r1 → create   r2 → repeat
r1 → print    r2 → create
r1 → repeat   r2 → repeat
r1 → repeat   r2 → print
```

For the last case, printing the results obtained by actions `g2` and `g3` will be organized in all positions of the regions `G1` reached by `g1` (which can be repeated by rule `r1` at the higher level).

3.3.2.2 *In branching operations*

Other rules may allow and supervise branching in space, with different branches (separated by comma) developing from the same positions in space (as follows and also shown in Figure 3.8, where `r1` is used to control two

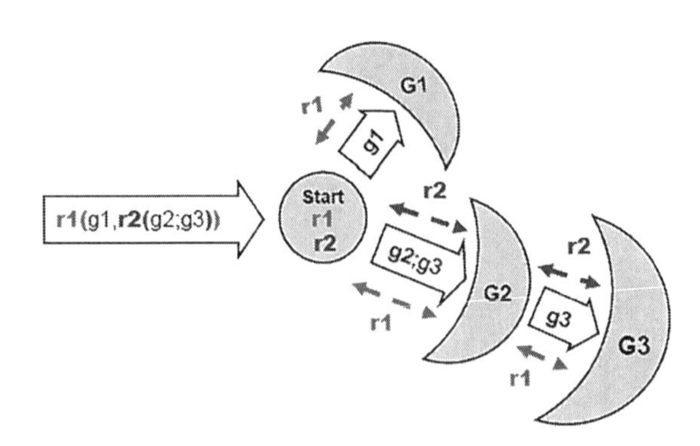

Figure 3.8 Branching under control rule.

branches, and r2 to manage the sequence of operations belonging to the second branch).

```
r1(g1, r2(g2; g3))
```

Possible meaning of r1 and r2 may be as follows, with both rules activated in the Start position.

```
r1 → or    r2 → create
```

Rule r1, activating two branches in any order or in parallel from the same Start position, selects the first one in time replying with a positive termination result, considering its space locations and data obtained there as the final result, while ignoring all achievements of another branch. Rule r2 covers the sequence of operations g2 and g3 of the second branch, supplying them with creative power during space navigation (will be completed and accepted if classified as the resultant branch by r1).

```
r1 → if    r2 → print
```

Rule r1 first launches branch g1, and only if and after it terminates with a positive result, activates the second branch with g2 and g3 embraced by rule r2, which organizes printing their final results in the Start position with staying there (the latter considered as holding the final result). If g1 results with failure, the final stay will be in the Start too but without any new result.

```
r1 → and   r2 → repeat
```

Rule r1 activates the two braches in any order or in parallel, and only after both branches reply with their final success, r1 can confirm the whole success of this mission (with r2 organizing repetitive development of the sequence of two embraced operations, which will eventually terminate). The successful positions reached in space by both branches will be considered as holding the final results of the scenario, with subsequent staying in them indefinitely. If any branch replies with failure, the development of the second branch will be terminated as soon as possible, as not needed any more. After r1 fails as a whole, there will be no position to stay in space further, with Start just abandoned.

We can also consider the development of operations g2 and g3 not in a sequence but in branching mode too, as follows, also depicted in Figure 3.9, with both rules activated in the Start position, where r1 is embracing r2 and all operations it coordinates, i.e., g2 and g3, as follows.

```
r1(g1, r2(g2, g3))
```

Possible examples of combinations of these rules:

```
r1 → or    r2 → or
r1 → if    r2 → and
r1 → and   r2 → or
r1 → and   r2 → and
```

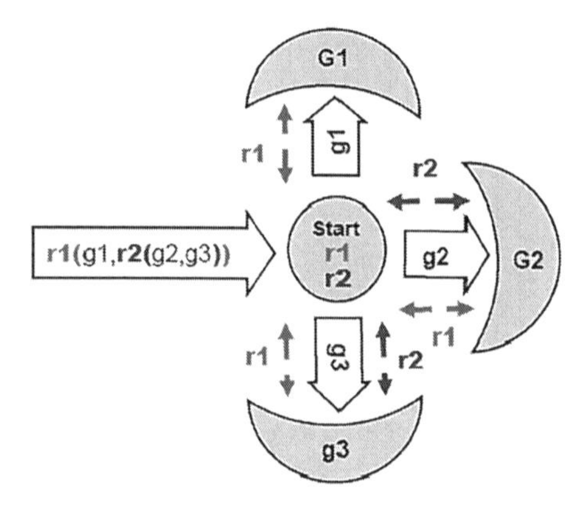

Figure 3.9 Nested branching.

3.3.3 Recursive hierarchy of scenarios

In Figures 3.2–3.9 we had in mind that operations g1, g2, g3 could be of any complexity. This means that each one may itself be substituted by the whole scenarios like the ones shown in all these figures, and for which each operation could be substituted by any scenario again, and so on, with such recursion potentially available to any depth. If, for example, to substitute operation g2 of the previous scenario r1(g1, r2(g2, g3)) shown in Figure 3.9 with the scenario (g4; r3(g5, g6)) enclosed in parentheses as a whole unit, we will receive the detailed combined scenario as follows, also shown in Figure 3.10:

```
r1(g1, r2((g4; r3(g5, g6)), g3))
```

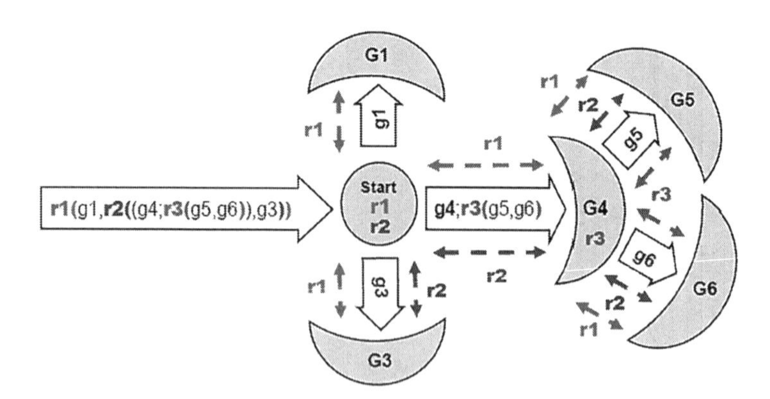

Figure 3.10 Scenario recursive extension.

3.3.4 Treating any operations as rules too

3.3.4.1 Collecting data for local processing

We may also consider all mentioned-above operations `gi` as rules which can have certain parameters as local constants, variables, or arbitrarily remote items. For example, we may have a rule providing a `sum` of values, like:

```
sum(g1, g2)
```

The rule's parameters, represented by operations `g1` and `g2` (which can be treated as rules too), may be having these values directly or may first need obtaining them remotely after the network search of any complexity and depth, as shown in Figure 3.11. The result will be obtained in the point where the rule started, i.e., `Start`, which will also be the final space point on the whole rule from which any further developments may take place (with the reached positions `G1` and `G2` subsequently abandoned).

Further development after this rule may be as follows, also shown in Figure 3.12.

```
sum(g1, g2); g3; g4
```

3.3.4.2 Local processing but leaving results remotely

Instead of the rule summing values we may have any other one like, for example, finding maximum from the values received by `g1` and `g2`, which will also produce the result in the rule `Start` position, also representing the final space position. But another rule like `maxdestination` (or `mxd` for short) may find the maximum value as before, but leave this value in the space position where it was originally found, with declaring this space position as the final one on the rule like, say, `G2` in Figure 3.13.

From this position `G2`, which may be remote, any further developments should take place (while abandoning `G1`), as follows and in Figure 3.14.

```
mxd(g1, g2); g3; g4
```

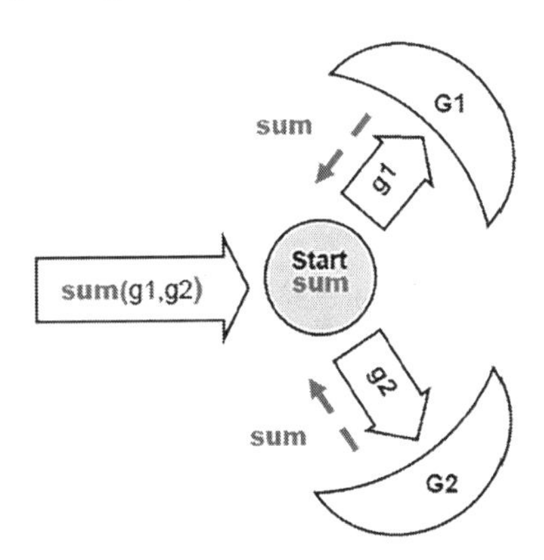

Figure 3.11 Expressing the summation operation by a rule.

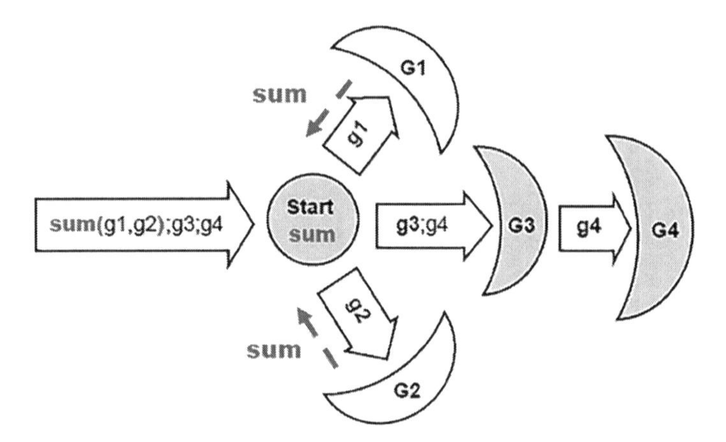

Figure 3.12 Example of further development from the rule's Start position.

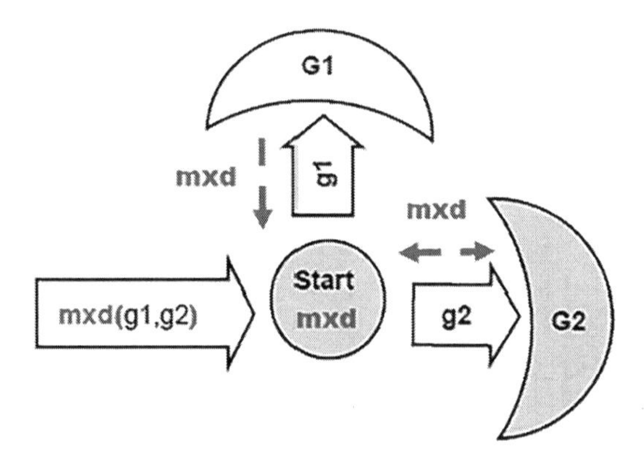

Figure 3.13 Example of maximum destination rule.

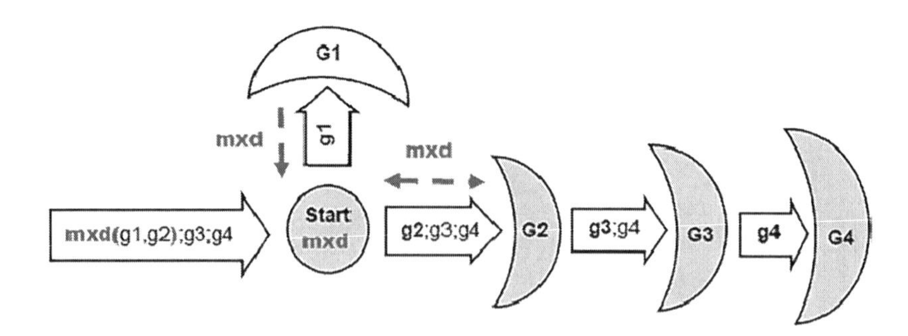

Figure 3.14 Example of further development from the G2 position.

3.3.5 Expressing sequences of operations by a rule too

We have shown only very few examples for structuring of spatial scenarios and rules coordinating their collections, with using semicolon to separate succeeding each other operations, and comma for separating branches starting from the same points. We may also have the same and unique separator for any collections of actions, embracing their sequences by rule `advance` (or `ad`) with comma as separator too, as follows (see also Figure 3.15, which can be used instead of Figure 3.3).

```
g1; g2; g3 → ad(g1, g2, g3)
```

3.3.6 The resultant unified recursive syntax of SG scenarios

The options discussed above and possibilities of structuring distributed scenarios with multiple operations allow us to obtain *a clear uniformity* of all possible spatial scenarios by the SG model with the use of recursion, as shown in Figure 3.16. Where the overall SG scenario is called *grasp*, syntactic categories are in italics, vertical bar separates alternatives, and the part in braces indicates zero or more repetitions using comma as a delimiter.

In general, the Spatial Grasp model under the top syntax of Figure 3.16 is *much more diverse, complex, and advanced* than what was shown by the restricted number of simple examples above, with capability of dynamic covering and matching of any distributed spaces and returning the obtained results and control states whatever remote and multiple they might be. It also allows us to make any decisions for the further space navigation, creates dynamic operational infrastructures capable of solving any distributed problems, and effectively implements, mimics, or simulates any other models and approaches (Petri nets and neural nets including), and so on.

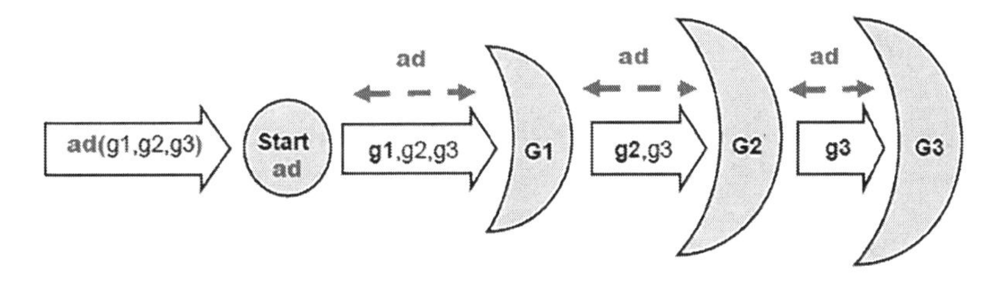

Figure 3.15 Using the rule to represent a sequence of operations.

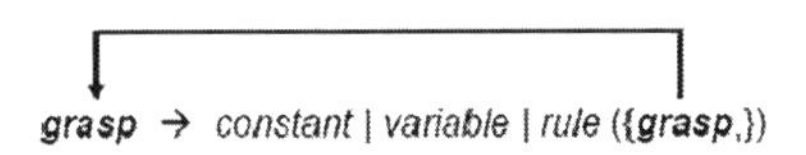

Figure 3.16 Universal recursive representation of the SG model.

3.3.6.1 Some possible flowchart extensions

SG has different types of variables that propagate when navigating distributed spaces, which together with returning remote results and data exchanges between different locations may be explicitly mentioned in spatiocharts too. An example of extension of the chart of Figure 3.15 with such DFE (Data Flow and Exchange) capability is shown in Figure 3.17.

Within the philosophy proclaimed in this book, even if the sets of positions `Gi` reached in physical or virtual spaces happen to be equal or included into the sets of previous positions `Gi-1`, and the operations `gi` remain in nodes belonging to `Gi-1`, we formally consider `Gi` positions as totally individual and new, and `gi` as organizing real propagation in space. But with certain clarification, if needed, we may distinguish operations and the reached regions which remain physically or virtually in the same space locations as the previous ones, as in the example below:

```
g1 → hop('Peter')
g2 → R = 10
g3 → move(x55_y88)
```

by just using dashed contours in charts as in Figure 3.18, where `g2`, `G1`, and `G2` may be staying in the same positions in space.

In another example:

```
R = 15; R = R + 10; move(x55_y88) → g1; g2; g3
```

we may have `Start`, `G1`, and `G2`, also operations `g1` and `g2` staying in the same world position `Start`, as in Figure 3.19.

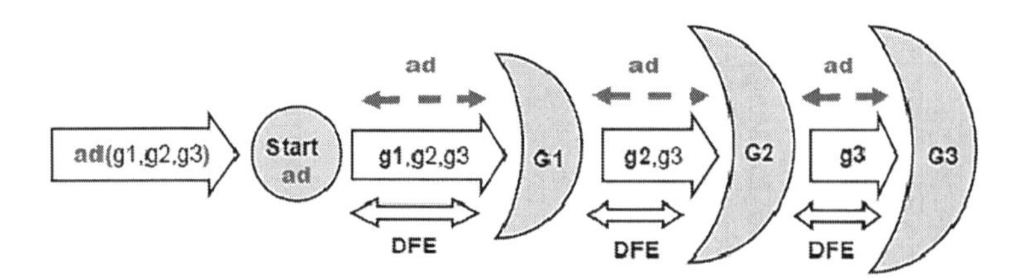

Figure 3.17 Extension of a spatiochart with data flow and exchange capability.

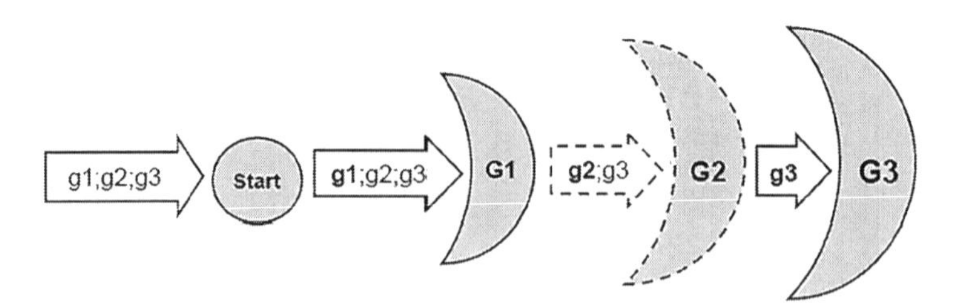

Figure 3.18 Distinguishing constructs remaining in the same physical positions.

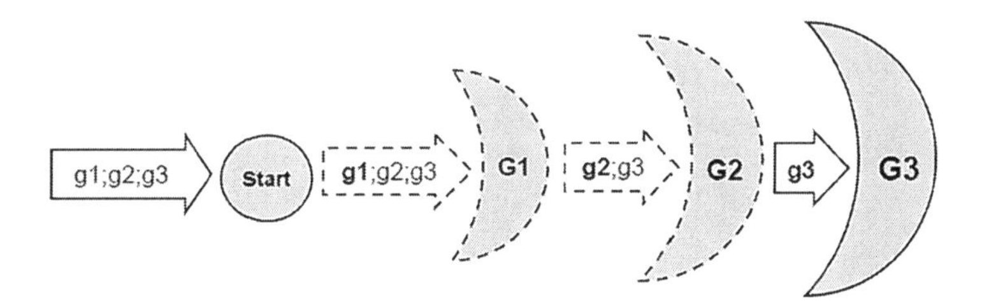

Figure 3.19 Another example with staying in the same physical positions.

But we will be using such dashed visualization options only if absolutely needed to explain details of spatial interpretation of SGL scenarios, which do not influence the general philosophy and semantics of SGL, also the nature of its spatiocharts.

3.3.6.2 Some scenario simplifications

Written SG scenarios, depending on implementation, may also use for simplification and shortening some constructs of traditional languages, like, for example, instead of writing: `assign(R, sum(R, 10))` to have more convenient: `R = R + 10`. Also, to use traditional semicolon for separation of the following each other constructs without embracing them with a rule, or separate branches just by comma without embracing them with a rule too if they are fully independent of each other, with comma being superior to semicolon in the expressions, as follows:

```
g1; (g2, g3, g4); (g5, g6) → g1; g2, g3, g4; g5, g6
```

In the next case, however, using parentheses will be absolutely important:

```
(g1; g2; g3), g4, (g5; g6)
```

But with any such simplifications, the general organizational structure of the SG model will always follow the one shown in Figure 3.16.

3.4 SPATIAL GRASP TECHNOLOGY (SGT) BASICS

Within Spatial Grasp Technology (SGT), a high-level scenario for any task to be performed in a distributed world is represented as an active self-evolving pattern rather than a traditional program, sequential or parallel one. This pattern, written in a high-level Spatial Grasp Language (SGL) and expressing top semantics of the problem to be solved, can start from any point of the world. Then it spatially propagates, replicates, modifies, covers,

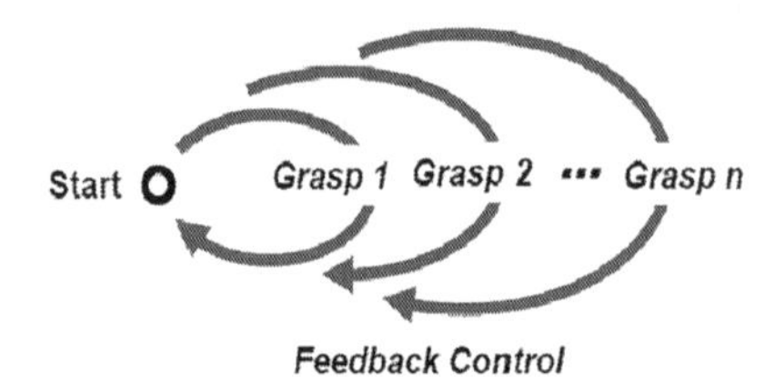

Figure 3.20 Controlled wavelike coverage and conquest of distributed spaces by SGT.

and matches the distributed world in a parallel wavelike mode, while echoing the reached control states and data found or obtained for making decisions at higher levels and further space navigation, as symbolically shown in Figure 3.20.

Many spatial processes in SGL can start any time and in any place, cooperating or competing with each other, depending on applications. The self-spreading and self-matching SGL patterns-scenarios can create active spatial infrastructures covering any regions. These infrastructures can effectively support or express distributed knowledge bases, advanced command and control, situation awareness, autonomous and collective decisions, as well as any existing or hypothetical computational and/or control models, systems, and solutions.

3.4.1 The Spatial Grasp Language (SGL)

The above-mentioned and many other SG model capabilities can be expressed by the recursive high level Spatial Grasp Language (SGL) in which all spatial scenarios are represented, with its top level syntax following (also graphically represented in Figure 3.21).

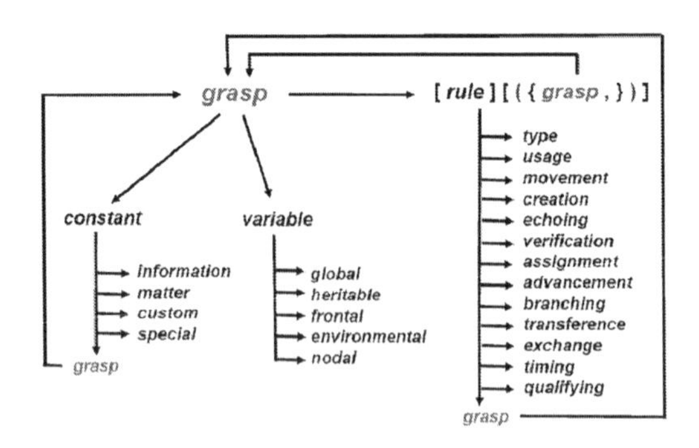

Figure 3.21 SGL recursive syntax.

```
grasp    → constant | variable | rule [({ grasp,})]
constant → information | matter | custom | special | grasp
variable → global | heritable | frontal | nodal | environmental
rule     → type | usage | movement | creation | echoing |
           verification | assignment | advancement | branching |
           transference | exchange | timing | qualifying | grasp
```

We will be briefing below main features of SGL and its distributed networked implementation.

3.4.2 The worlds SGT operates with

SGT allows us to directly operate with the following world representations: *Physical World* (PW), considered as continuous and infinite, where each point can be identified and accessed by physical coordinates; *Virtual World* (VW), which is discrete and consists of nodes and semantic links between them; and *Executive world* (EW) consisting of active "doers" with communication possibilities between them. Different kinds of combination of these worlds can also be possible within the same formalism, as follows: *Virtual-Physical World* (VPW) where individually named VW nodes can associate with coordinates of certain PW points or any of its regions; *Virtual-Execution World* (VEW), where doer nodes may have special names assigned to them and semantic relations in between, similarly to pure VW nodes; *Execution-Physical World* (EPW) can have doer nodes associated with certain PW coordinates; and *Virtual-Execution-Physical World* (VEPW) combining all features of the previous cases.

3.4.3 SGL constants

Different types of constants can be used in SGL. *Information* can be represented by *number* in its traditional syntax or by any *string* of characters which may also reflect a full scenario text or its part suitable for direct automatic interpretation; it can correspond to physical *matter* (physical objects including) too. A repertoire of self-identifiable *special* and *custom* constants can be used as standard parameters (or modifiers) in different language rules. Constants can also be compound ones, using recursive *grasp* definition in SGL syntax, which allows us to represent any nested and hierarchical structures consisting of multiple (elementary or compound again) objects.

3.4.4 SGL variables

Spatial variables, stationary or mobile, which can be used in fully distributed physical, virtual, or executive environments, are effectively serving multiple cooperative processes under the unified control. These are: *Global variables* (most expensive), which can serve any SGL scenarios and be shared by them and by their different branches; *Heritable variables* appearing within a scenario step and serving all subsequent, descendent steps; *Frontal variables* serving and accompanying the scenario evolution, being transferred between subsequent steps; *Environmental variables* allowing us to access, analyze, and possibly change different features of physical, virtual, and executive words during their navigation; and finally, *Nodal variables* as a property of the world positions reached by scenarios and shared with other scenarios in same positions.

3.4.5 SGL rules

SGL rules, capable of representing any actions or decisions, belong to the following main categories: (a) hierarchical fusion and return of potentially remote data; (b) distributed control, sequential and/or parallel, in both breadth and depth of the scenario evolution; (c) a variety of special contexts detailing navigation in space, also clarifying character and peculiarities of the embraced operations and decisions; (d) type and sense of a value or its chosen usage for guiding automatic language interpretation; and (e) individual or massive creation, modification, or removal of nodes and connecting links in distributed knowledge networks, allowing us to effectively work with arbitrary knowledge structures. All rules are pursuing the same unified ideology and organizational scheme, as follows: (1) they start from a certain world position, being initially linked to it; (2) perform or control the needed operations in a distributed space, which may be branching, stepwise, parallel, and arbitrarily complex, also local and remote; and (3) produce or supervise concluding results of the scenario embraced, expressed by control states and values in different points.

3.4.6 Control states

The following control states can appear after completion of different scenario steps. Indicating local progress or failure, they can be used for effective control of multiple distributed processes with proper decisions at different levels. These states are: *thru*—reflects full *success* of the current scenario branch with capability of further development; *done*—indicates success of the current scenario step with its planned termination; *fail*—indicates nonrevocable failure of the current branch and no possibility of further development from the location reached; and *fatal*—reporting terminal failure with nonlocal effect, while triggering massive abortion of all currently evolving scenario processes and removal of associated temporary data with them. These control states, appearing in different branches of parallel and distributed scenario at bottom levels, can be used to obtain generalized control states at higher levels, up to the whole scenario, in order to make proper decisions for the further scenario evolution.

3.4.7 How SGL scenarios evolve

More details on how SGL scenarios self-evolve in distributed environments.

a) SGL scenario is considered developing in *steps*, which can be *parallel*, with new steps produced on the basis of previous steps.
b) Any step, including the starting one, is always associated with a certain *point* or position of the world (i.e., physical, virtual, executive, or combined) from which the scenario (or its particular part, as there may be many parts working simultaneously) is currently developing.
c) Each step provides a resultant *value* (which may be single or multiple, also structured) representing information, matter, or both, and a resultant control *state* (as one of possible states ranging by their strength, shown later). This resultant state may be evaluated and issued in the step's starting point whereas local states can also be issued in the points reached by the step, which may be multiple.
d) Different scenario parts may evolve from the same points in *ordered*, *unordered*, or *parallel* manner, as independent or interdependent steps-branches.
e) Different scenario parts can also spatially *succeed* each other, with new parts evolving from final positions and results produced by the previous parts.
f) This potentially parallel and distributed scenario evolution may proceed in *synchronous* or *asynchronous* modes, and any of their combinations.
g) SGL operations and decisions in evolving scenario parts can use control states and values *returned* from other scenario parts however complex and remote they might be, thus combining *forward* and *backward* scenario evolution in distributed spaces.
h) Different steps from the same or different scenario parts may happen to be temporarily associated with the same, reached, world points while sharing persistent or provisional information in them.
i) Staying with world points, it is possible to *change* local parameters in them, whether physical or virtual, thus *impacting* the navigated worlds via these locations.
j) Scenarios navigating distributed spaces can *create arbitrarily distributed physical or virtual infrastructures* in them, which may operate on their own after becoming active, with or without additional external control. They can also be subsequently (or even during their creation) navigated, updated, and processed by same or other scenarios.
k) Overall organization of the world creation, navigation, coverage, modification, analysis, and processing can be provided by a variety of SGL rules which may be arbitrarily *nested*.
l) The evolving SGL scenario, as already mentioned, can *lose utilized parts* if not needed any more; it can also *self-modify* and *self-replicate* during space navigation, to adjust to the environments and optimize communications in distributed systems.

3.4.8 Networked SGL interpreter

3.4.8.1 General on SGL interpretation

Communicating Interpreters of SGL can be in arbitrary number of copies, say, up to millions and billions, which can be effectively integrated with any existing systems and communications, and their dynamic networks can represent powerful spatial engines capable of solving any problems in terrestrial and celestial environments. Such collective engines can simultaneously execute many cooperative or competitive tasks without any central resources or control, as symbolically depicted in Figure 3.22 (SGL interpreters just named U as universal computational and management nodes).

3.4.8.2 Some interpreter details

The main components of the SGL interpreter are shown in Figure 3.23.

The interpreter consists of a number of specialized *functional processors* (shown by rectangles) which are working with specific data structures. These include: Communication Processor, Control Processor, Navigation Processor, Parser, different Operation Processors, and special World Access

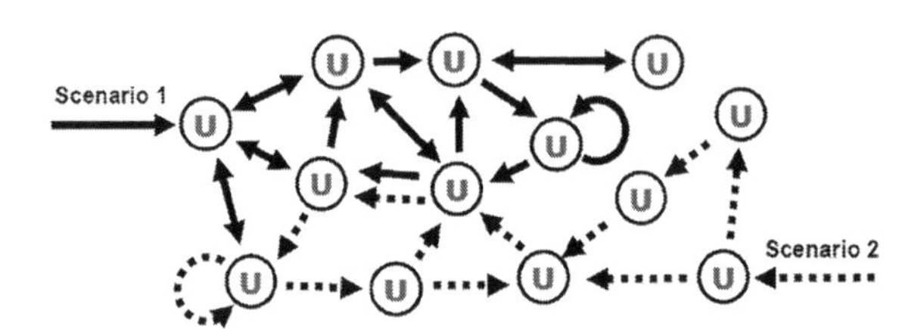

Figure 3.22 SGL interpretation networks as a global world computer.

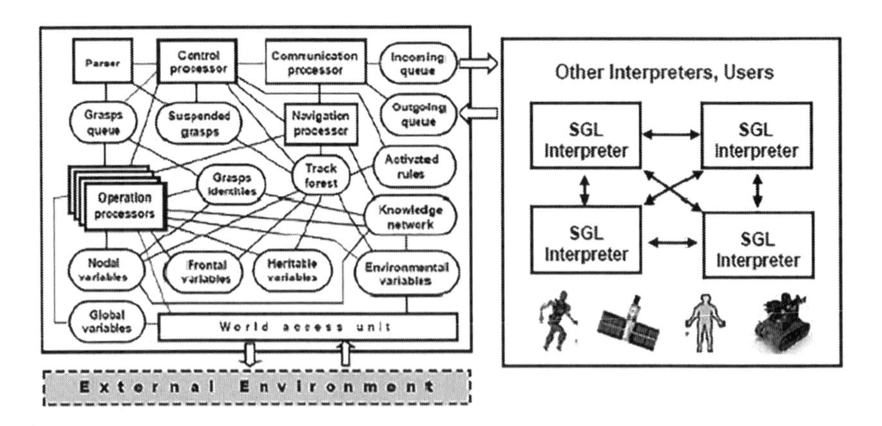

Figure 3.23 Main components of SGL interpreter and its networked organization.

Unit. Main *data structures* (shown by ovals) comprise: Grasps Queue, Suspended Grasps, Track Forest, Activated Rules, Knowledge Network, Grasps Identities, Heritable Variables, Fontal Variables, Nodal Variables, Environmental Variables, Global Variables, Incoming Queue, and Outgoing Queue.

Each interpreter can support and process multiple SGL scenario codes which happen to be in its responsibility at different moments of time. Integrated with any distributed systems, the interpretation network allows us to form a spatial world computer with practically unlimited power for simulation and management of the universe.

3.4.8.3 Spatial track system

As both backbone and nerve system of the distributed interpreter, its self-optimizing *spatial track system* provides hierarchical command and control as well as remote data and code access. It also supports spatial variables and merges distributed control states for decisions at different organizational levels. The track infrastructure is automatically distributed between active components (humans, robots, computers, smartphones, satellites, etc.) during scenario self-spreading in distributed environments. The main track components are shown in Figure 3.24, to be used in the subsequent figures.

- Forward grasping

  In the forward process, as in Figure 3.25, the next steps of scenario development are forming new 3track nodes connected to the previous nodes by track links. Reflecting the history of scenario evolution, this growing track structure is effectively supporting heritable, nodal, and

Figure 3.24 Main track components.

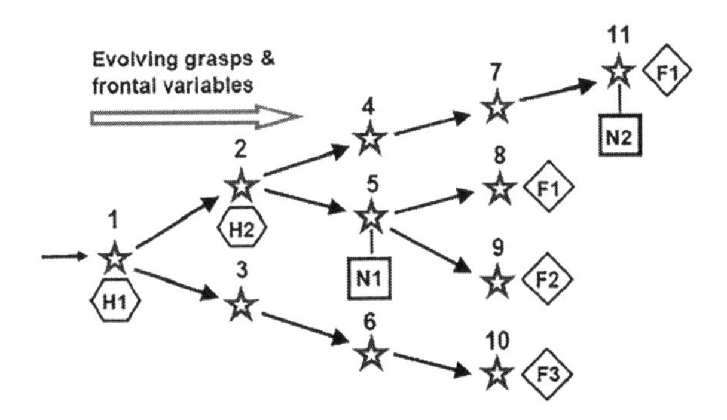

Figure 3.25 Forward world grasping.

frontal variables, also activated rules and suspended grasps (associated with proper track nodes).

- **Echoing**

  After completing the forward stage of SGL scenario, the track system can return to the starting track node, the generalized control state based on termination states in all fringe nodes, as in Figure 3.26, also marking the passed track links with the states returned via them. The track system, on the request of higher-level scenario rules, can also collect local data obtained at its fringe nodes and merge them into a resultant list of values echoed to the starting node. The track echoing process also optimizes the track system by deleting already used and not needed any more items.

- **Further forward development**

  The echo-modified and optimized track system can route further grasps to the world positions reached by the previous grasps and defined by fringe track nodes having state thru, as in Figure 3.27.

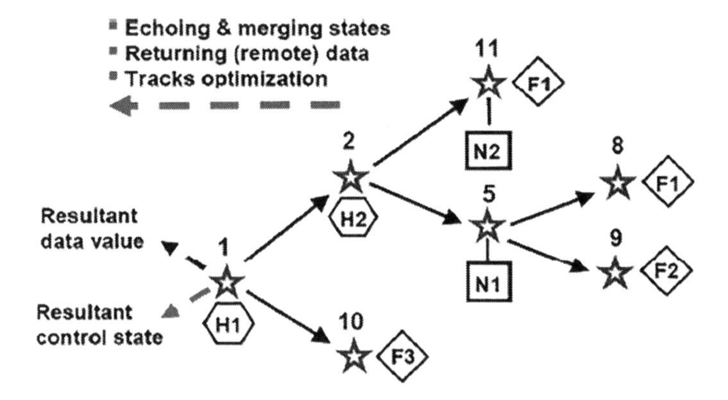

Figure 3.26 Echoing and tracks optimization.

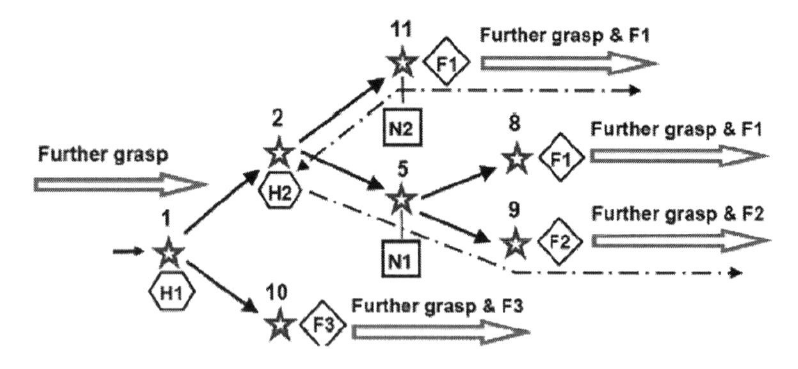

Figure 3.27 Further world grasping via optimized tracks.

Heritable variables created in certain track nodes can also be accessed from the subsequent nodes in the track system for both reading and writing operations.

Detailed information on SGT, SGL, and its networked interpreter, also solving numerous problems from very different classes under such approach, can be obtained from many existing publications, including [7–13], also just by *spatial grasp* in google.com

3.5 CONCLUSIONS

This chapter concludes with the experience obtained by many years of dealing with distributed networked systems and their applications in very different areas; the development of distributed control technology with its Spatial Grasp Language has also been summarized and generalized. This generalization has been made in the form of radically new computational and control model as an extension of traditional concepts of algorithm and its flowcharts, allowing us to *directly operate in combined terrestrial and celestial environments*. The introduced new concept of flowcharts, called *spatiocharts*, also helps us to exhibit, analyze, and develop complex solutions for distributed systems operating on Earth and in outer space. Instead of representing distributed systems as a collection of parts or agents exchanging messages, we have their integral and holistic solutions as *self-evolving and self-matching spatial patterns* on a high semantic level, which can often simplify and dramatically reduce the global management code up to a hundred times in comparison with other approaches and languages. The latest applications of the discussed spatial model and the resultant control technology can be found in [14–31], with many of them also presented in detail in the subsequent chapters.

REFERENCES

1. Algorithm. https://en.wikipedia.org/wiki/Algorithm
2. Definition of Algorithm. Merriam-Webster Online Dictionary. https://www.merriam-webster.com/dictionary/algorithm
3. M.T. Goodrich, R. Tamassia, Algorithm Design: Foundations, Analysis, and Internet Examples, John Wiley & Sons, Inc. (2002). Archived from the original on April 28, 2015, retrieved June 14, 2018. https://web.archive.org/web/20150428201622/http:/ww3.algorithmdesign.net/ch00-front.html
4. Heuristic. https://en.wikipedia.org/wiki/Heuristic
5. Flowchart. https://en.wikipedia.org/wiki/Flowchart
6. A. Lynch, Flow Chart Design - How to design a good flowchart, 04/22/2021. https://www.edrawsoft.com/flowchart-design.html?gclid=Cj0KCQjw5auGBhDEARIsAFyNm9H7cWvesxjTsposJTaD890zBspEUDA18dHi_9R_GChiyzitwgpYCu4aAjutEALw_wcB

7. P.S. Sapaty, Symbiosis of Real and Simulated Worlds under Spatial Grasp Technology, Springer, 2021. 251 p.
8. P.S. Sapaty, Complexity in International Security: A Holistic Spatial Approach, Emerald Publishing, 2019. 160 p.
9. P.S. Sapaty, Holistic Analysis and Management of Distributed Social Systems, Springer, 2018. 234 p.
10. P.S. Sapaty, Managing Distributed Dynamic Systems with Spatial Grasp Technology, Springer, 2017. 284 p.
11. P.S. Sapaty, Ruling Distributed Dynamic Worlds. New York: John Wiley & Sons, 2005. 255 p.
12. P.S. Sapaty, Mobile Processing in Distributed and Open Environments. New York: John Wiley & Sons, 1999. 410 p.
13. P.S. Sapaty, A distributed processing system, European Patent N 0389655, Publ. 10.11.93, European Patent Office. 35 p.
14. P.S. Sapaty, Global Network Management under Spatial Grasp Paradigm. *International Robotics & Automation Journal* 6(3) (2020), 134–148, 134. https://medcraveonline.com/IRATJ/IRATJ-06-00212.pdf
15. P. S. Sapaty, Global Network Management under Spatial Grasp Paradigm, *Global Journal of Researches in Engineering: J General Engineering* 20(5) (2020), 58–81. https://globaljournals.org/GJRE_Volume20/6-Global-Network-Management.pdf
16. P.S. Sapaty, Advanced terrestrial and celestial missions under spatial grasp technology, *Aeronautics and Aerospace Open Access Journal* 4(3) (2020). https://medcraveonline.com/AAOAJ/AAOAJ-04-00110.pdf
17. P.S. Sapaty, Spatial Management of Distributed Social Systems, *Journal of Computer Science Research* 2(3) (2020, July). https://ojs.bilpublishing.com/index.php/jcsr/article/view/2077/pdf
18. P.S. Sapaty, Towards Global Nanosystems under High-level Networking Technology, *Acta Scientific Computer Sciences* 2(8) (2020). https://www.actascientific.com/ASCS/pdf/ASCS-02-0051.pdf
19. P.S. Sapaty, Symbiosis of Distributed Simulation and Control under Spatial Grasp Technology, *SSRG International Journal of Mobile Computing and Application (IJMCA)* 7(2) (2020, May–August). http://www.internationaljournalssrg.org/IJMCA/2020/Volume7-Issue2/IJMCA-V7I2P101.pdf
20. P.S. Sapaty, Global Network Management under Spatial Grasp Paradigm, *International Robotics & Automation Journal* 6(3) (2020). https://medcraveonline.com/IRATJ/IRATJ-06-00212.pdf
21. P.S. Sapaty, Global Network Management under Spatial Grasp Paradigm, *Global Journal of Researches in Engineering: J General Engineering* 20(5) (2020) Version 1.0. https://globaljournals.org/GJRE_Volume20/6-Global-Network-Management.pdf
22. P.S. Sapaty, Symbiosis of Virtual and Physical Worlds under Spatial Grasp Technology, *Journal of Computer Science & Systems Biology* 13(6) (2020). https://www.hilarispublisher.com/open-access/symbiosis-of-virtual-and-physical-worlds-under-spatial-grasp-technology.pdf
23. P.S. Sapaty, Symbiosis of Real and Simulated Worlds Under Global Awareness and Consciousness, *The Science of Consciousness | TSC* 2020. https://eagle.sbs.arizona.edu/sc/report_poster_detail.php?abs=3696

24. P.S. Sapaty, Spatial Grasp as a Model for Space-based Control and Management Systems, *Mathematical Machines and Systems* 1 (2021), 135–138. http://www.immsp.kiev.ua/publications/articles/2021/2021_1/Sapaty_book_1_2021.pdf
25. P.S. Sapaty, Managing multiple satellite architectures by spatial grasp technology, *Mathematical Machines and Systems* 1 (2021), 3–16. http://www.immsp.kiev.ua/publications/eng/2021_1/
26. P.S. Sapaty, Spatial Management of Large Constellations of Small Satellites, *Mathematical Machines and Systems* 2 (2021). http://www.immsp.kiev.ua/publications/articles/2021/2021_2/02_21_Sapaty.pdf
27. P.S. Sapaty, Global Management of Space Debris Removal under Spatial Grasp Technology, *Acta Scientific Computer Sciences* 3(7) (2021, July). https://www.actascientific.com/ASCS/pdf/ASCS-03-0135.pdf
28. P.S. Sapaty, Space Debris Removal under Spatial Grasp Technology, *Network and Communication Technologies* 6(1) (2021). https://www.ccsenet.org/journal/index.php/nct/article/view/0/45486
29. P.S. Sapaty, Spatial Grasp Model for Management of Dynamic Distributed Systems, *Acta Scientific Computer Sciences* 3(9) (2021). https://www.actascientific.com/ASCS/pdf/ASCS-03-0170.pdf
30. P.S. Sapaty, Spatial Grasp Model for Dynamic Distributed Systems, *Mathematical Machines and Systems* 3 (2021). http://www.immsp.kiev.ua/publications/articles/2021/2021_3/03_21_Sapaty.pdf
31. P.S. Sapaty, Development of Space-based Distributed Systems under Spatial Grasp Technology, *Mathematical Machines and Systems* 4 (2021).

Chapter 4

Spatial Grasp Language (SGL)

4.1 INTRODUCTION

This chapter is about the Spatial Grasp Language (SGL) as the basic element of the Spatial Grasp Technology (SGT) described in Chapter 3, and in more detail in [1–6], in which a high-level scenario for any task to be performed in a distributed world is represented as an active space-conquering and self-evolving pattern rather than a traditional program, sequential or parallel one. This pattern, written in SGL and expressing top semantics of the problem to be solved, can start from any point of the world. Then it spatially propagates, replicates, modifies, covers, and matches the distributed world in a parallel wavelike mode, while echoing the reached control states and data found or obtained for making decisions at higher levels and further space navigation, as symbolically shown in Figure 4.1.

Many spatial processes in SGL can start any time and in any place, cooperating or competing with each other, depending on applications. The self-spreading and self-matching SGL patterns-scenarios can create active spatial infrastructures covering any regions. These infrastructures can effectively support or express distributed knowledge bases, advanced command and control, situation awareness, autonomous and collective decisions, as well as any existing or hypothetical computational and/or control models, systems, and solutions. SGL has a deep recursive structure with its parallel spatial scenarios called *grasps* and universal operational and control constructs as *rules* (braces identifying repetition):

```
grasp → constant | variable | rule ({ grasp,})
```

The chapter offers full details of the latest SGL version suitable for dealing with very large distributed terrestrial and celestial systems. It describes different types of constants capable of representing information, physical matter, or both, and specific types of variables, called spatial, as operating in fully distributed spaces and being mobile themselves when serving spreading algorithms. It also provides a full repertoire of the language main constructs called rules, which can be arbitrarily nested and may carry different navigation, creative, processing, control, and other loads. The rules equally operate

DOI: 10.1201/9781003230090-4

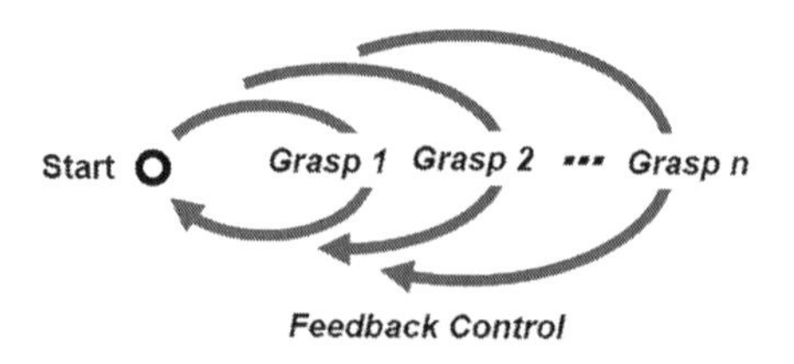

Figure 4.1 Controlled wavelike coverage and conquest of distributed spaces by SGT.

with local and remote values, readily process both physical matter/objects and distributed networked knowledge, and can be used for effective expression of active graph-based spatial patterns, which are navigating, invading, matching, processing, conquering, and changing distributed environments. Elementary programming examples in SGL are provided too. Details on the previous SGL versions and numerous examples of programming in them can be found in the previous books [1–6], past publications (with more in Chapter 1) [7–22], and the latest ones [23–41].

The rest of this chapter is organized as follows. *Section 4.2* provides full SGL syntax and explains general issues of its organization. *Section 4.3* describes SGL constants, which include information, physical matter, special and custom constants, as well as arbitrary complex or compound ones. *Section 4.4* provides the repertoire of SGL spatial variables, which includes global, heritable, frontal, nodal, and environmental variables. *Section 4.5* describes and explains main SGL rules that provide such functionalities and operations as usage, movement, creation, echoing, verification, assignment, advancement, branching, transference, exchange, timing, qualification, and grasping. *Section 4.6* contains some examples of spatial scenarios in SGL, which include distributed network management, organization of integral human-robotic collectives, and simulation of worldwide spreading of and also fighting with malicious viruses. *Section 4.7* concludes the chapter.

4.2 FULL SGL SYNTAX AND GENERAL ISSUES

We are starting here with the SGL full syntax description, where syntactic categories are shown in italics, vertical bar separates alternatives, parts in braces indicate zero or more repetitions with a delimiter at the right if multiple, and constructs in brackets are optional. The remaining characters and words are the language symbols (including boldfaced braces).

```
grasp        →  constant | variable | [ rule ] [({
                grasp,})]
constant     →  information | matter | special | custom |
                grasp
information  →  string | scenario | number
string       →  '{character}'
```

```
scenario          →  {{character}}
number            →  [sign]{digit}[.{digit}[e[sign]{digit}]]
matter            →  "{character}"
special           →  thru | done | fail | fatal | infinite
                     | nil | any | all | other | allother |
                     current | passed | existing | neighbors |
                     direct | forward | backward | synchronous
                     | asynchronous | virtual | physical |
                     executive | engaged | vacant | firstcome
                     | unique
variable          →  global | heritable | frontal | nodal |
                     environmental
global            →  G{alphameric}
heritable         →  H{alphameric}
frontal           →  F{alphameric}
nodal             →  N{alphameric}
environmental     →   TYPE | NAME | CONTENT | ADDRESS |
                      QUALITIES |
                      WHERE | BACK | PREVIOUS | PREDECESSOR |
                      DOER |
                      RESOURCES | LINK | DIRECTION | WHEN |
                      TIME | STATE |
                      VALUE | IDENTITY | IN | OUT | STATUS
rule              →  type | usage | movement | creation |
                     echoing | verification |
                     assignment | advancement | branching |
                     transference |
                     exchange | timing | qualifying | grasp
 type             →  global | heritable | frontal | nodal |
                     environmental | matter | number | string |
                     scenario | constant | custom
usage             →  address | coordinate | content | index |
                     time | speed | name | place | center |
                     range |
                     doer | node | link | unit
movement          →  hop | hopfirst | hopforth | move | shift
                     | follow
creation          →  create | linkup | delete | unlink
echoing           →  state | rake | order | unit | unique |
                     sum |
                     count | first | last | min | max | random
                     | average |
                     sortup | sortdown | reverse | element
                     | position | fromto | add | subtract |
                     multiply | divide |
                     degree | separate | unite | attach |
                     append |
                     common | withdraw | increment | decrement
                     | access | invert | apply | location |
                     distance
```

```
verification    →  equal | nonequal | less | lessorequal |
                   more | moreorequal | bigger | smaller |
                   heavier |
                   lighter | longer | shorter | empty |
                   nonempty | belong | notbelong | intersect
                   | notintersect | yes | no
assignment      →  assign | assignpeers
advancement     →  advance | slide | repeat | align | fringe
branching       →  branch | sequence | parallel | if | or |
                   and |
                   or_sequence | or_parallel | and_sequence
                   | and_parallel |
                   choose | quickest | cycle | loop | sling
                   | whirl | split | replicate
transference    →  run | call
exchange        →  input | output | send | receive | emit |
                   get
timing          →  sleep | allowed
qualification   →  contain | release | trackless | free |
                   blind | quit |
                   abort | stay | lift | seize | exit
```

From this language definition, an SGL scenario, called *grasp*, supposedly applied in some point of the distributed space, can just be a *constant* directly providing the result to be associated with this point. It can be a *variable* whose content, assigned to it previously when staying in this or (remotely) in other space point (as variables may have non-local meaning and coverage), provides the result in the application point too. It can also be a *rule* (expressing certain action, control, description, or context) optionally accompanied with operands separated by comma (if multiple) and embraced in parentheses. These operands can be of any nature and complexity (including arbitrary scenarios themselves) and defined recursively as *grasp*, i.e., can be constants, variables or any rules with operands (i.e., as grasps again), and so on.

Rules, starting in some world points, can organize navigation of the world sequentially, in parallel or any combinations thereof. They can result in staying in the same application point or can cause movement to other world points with obtained results to be left there, as in the rule's final points. Such results can also be collected, processed, and returned to the rule's starting point, the latter serving as the final one on this rule. The final world points reached after the rule invocation can themselves become starting ones for other rules. The rules, due to recursive language organization, can form arbitrary operational and control infrastructures expressing any sequential, parallel, hierarchical, centralized, localized, mixed, and up to fully decentralized and distributed algorithms. These algorithms, called *spatial*, can effectively operate *in*, *with*, *under*, *in between*, *over*, and *instead of* (as for simulation) large, dynamic, and heterogeneous spaces, which can be physical, virtual, management, command and control, or combined.

Let us describe and analyze SGL features in more detail.

4.3 SGL CONSTANTS

The definition of constants, which can be of different types, is as follows:

```
constant → information | matter | special | custom | grasp
```

Constants can be self-identifiable by the way they are written, as follows in this section, or defined by special rules embracing them for arbitrary textual representations, as shown later.

4.3.1 Information

Information constants can be of the following categories:

```
information → string | scenario | number | custom
```

- **String**
  A string can be represented in most general way as a sequence of characters embraced by opening-closing single quotation marks:

  ```
  string → '{character}'
  ```

  This sequence should not contain other single quotes inside unless they appear in opening-closing pairs, with such nesting allowed to any depth. If single words representing information are not intersecting with other language constructs, the quotes around them can be omitted.
- **Scenario**
  Another string representation may be in the form of explicit SGL scenario body:

  ```
  scenario → {{character}}
  ```

  For this case a sequence of characters should be placed into opening-closing curly brackets, or braces **{}**, shown here in bold (to distinguish from braces used for textual repetition in the language syntax), which can be used inside the string and nested in pairs too. Braces will indicate the text as a potential *scenario* code which should be optimized *before* its any usage. This may involve removing unnecessary spaces, substituting names of rules by their allowed shortcuts, or adjusting to the standard SGL syntax after using constructs typical to other programming languages for convenience (like semicolons as separators instead of sequencing rules). If single quotes embrace SGL texts to be used as an executable scenario, such code optimization will have to be done each time *during* its interpretation, not before, with the original text remaining intact.

- **Number**
  Number can be represented in a standard way, similar to traditional programming languages, generally in the following form (with brackets identifying optional parts and braces the repeating characters).

  ```
  [sign]{digit}[.{digit}[E[sign]{digit}]]
  ```

  Numbers can also use words instead of digits and accompanying characters like sign and dot (with underscore as separator if more then one word needed to represent them).

4.3.2 Physical matter

Physical *matter* (physical objects including) in the most general way can be reflected in SGL by a sequence of characters embraced by opening-closing double quotation marks:

```
matter → "{character}"
```

4.3.3 Special constants

Special or reserved constants (as free standing words, without quotes) may be used as standard parameters (or modifiers) in different language rules, with frequently engaged ones following.

```
special → thru | done | fail | fatal | infinite | nil |
any | all | other | allother | current | passed | existing
| neighbors | direct | forward | backward | neutral |
synchronous | asynchronous | virtual | physical | executive
| engaged |
vacant | firstcome | unique
```

Let us explain their possible meanings and applications.

thru—indicates (or artificially sets up) control state of the scenario in the current world point as an *absolute success* with possibility of further scenario evolution from this particular point. This does not influence scenario developments in other world points.

done—indicates (or sets artificially up) control state as a *successful termination* in the current world point, with blocking further scenario development from this particular point. Not influencing scenario developments in other world points.

fail—indicates (or artificially sets up) scenario control state as *failure* in the current world point, without possibility of further scenario development from this particular point. Not influencing scenario developments in other world points.

fatal—indicates (or sets up artificially) control state as nonlocal *fatal failure* starting in the current world point and causing massive termination and removal of all active distributed processes with related local data in this and other world points reached by the same scenario (which may have parallel branches). The destructive influence of this state may be contained at higher levels by special rules explained later.

infinite—indicates infinitely large value.

nil—indicates no value at all.

any, all, other, allother—stating that *any one*, *all* (the current one including), *any other*, or *all other* (the current one excluding in the last two cases) elements can be considered and used by some rule.

current—refers to the current element (like node) only, for its further consideration or reentering (possibly, with proper conditions).

passed—informing that the mentioned elements (like world nodes) have already been passed by the current scenario branch on its way to the current point. These elements may in some cases be accessed easier by backwarding via the SGL history-based distributed control than by a global search just by name.

existing—hinting that world nodes with given names which are currently under consideration already exist and should not be created again (i.e., duplicated).

neighbors—stating that the nodes to be accessed are among direct neighbors of the current node, i.e., located within a single hop from it via the existing links.

direct—stating that the mentioned nodes should be accessed or created from the current node directly, without consideration of possible semantic links to them (even if such links already exist, as in the case of accessing neighboring nodes).

forward, backward, neutral—allowing us to move from the current node via existing links along, against, or regardless their orientations (ignored when dealing with nonoriented links, which can always be traversed in both directions).

synchronous, asynchronous—a modifier setting synchronous or asynchronous mode of operations induced by different rules.

virtual, physical, executive—indicating or setting the type of a node the scenario is currently dealing with (the node can also be of a combined type, having more than one such indicator, with maximum three).

engaged, vacant—indicating or setting the state of a resource the current scenario is dealing with (like, human, robot, or any other physical, virtual, or combined world node).

firstcome—allows the current scenario with its unique identity to enter the world nodes only first time (the capability based on internal language node marking mechanisms, which can be used for different purposes, including blocking of unplanned or unwanted cycling).

unique– allows the return of only unique elements received from the embraced scenario final positions while omitting duplicates (the returned results should form a list without repeating elements).

Any other parameters or modifiers can be used in SGL scenarios too, say, as general constants (strings or numbers) if their repertoire and meanings are not covered by the abovementioned examples.

4.3.4 Custom constants

Other self-identifiable, or *custom*, constants can be incorporated to be directly processed by updated SGL, if they do not conflict with the language syntax; otherwise they should be declared by special rules. They can represent standing without quotes or braces information and physical matter substances or objects, as well as their combinations.

4.3.5 Compound constants

Constants can also be compound ones, using the recursive *grasp* definition in SGL syntax, which allows us to represent nested hierarchical structures consisting of multiple (elementary or compound again) objects. This, in particular for constants, can be expressed as:

```
constant → rule({ constant,})
```

Different SGL rules explained later may be used for such structuring, with more to be added for particular applications.

4.4 SGL VARIABLES

There are five types of SGL variables, called *spatial*, serving quite differently multiple cooperative processes in distributed virtual, physical, executive, and combined spaces, as follows:

```
variable → global | heritable | frontal | nodal |
environmental
```

Different types of variables can be self-identifiable by the way their names are written. Their names can also have any textual representations if explicitly declared by special rules explained later.

4.4.1 Global variables

This is the most expensive type of SGL variables, with their names starting with capital G and followed by arbitrary sequences of alphabetic letters and/or digits:

```
global → G{alphameric}
```

These variables can exist only in single copies with particular names, being common for both read and write operations to all processes of the same scenario, regardless of their physical or virtual distribution and world points they may cover. Global variables can be created by first assignment to them within any scenario branch and used afterwards by the entire scenario, including all its branches. They cease to exist only when removed explicitly or the whole scenario that created them terminates.

4.4.2 Heritable variables

The names of these variables should start with capital `H` if not defined by a special rule:

```
heritable → H{alphameric}
```

Heritable variables, being created by first assignment to them at some scenario development stage, are becoming common for read-write operations *for all subsequent* scenario operations (generally multiple, parallel, and distributed) evolving from this particular point and wherever there is space they may happen to be. This means that such variables are unique only within concrete hereditary scenario developments, to all their depth. The life time of these variables depends on the continuing activity of processes that can potentially inherit them, with their removal made explicitly or after all such processes terminate. Heritable variables can also model global variables if declared at the very beginning of the scenario starting from a single point, as all scenario developments can be using and sharing them afterwards.

4.4.3 Frontal variables

These are mobile-type variables with names starting with capital `F`, which are propagating in distributed spaces while keeping their contents on the forefronts of evolving scenarios:

```
frontal → F{alphameric}
```

Each of these variables is serving only the current scenario branch operating in the current world point. They cannot be shared with other branches evolving in the same or other world points, while always accompanying the scenario control. If the scenario splits into individual branches in the same world point or when moving to other points, these variables are replicated with the same names and contents and serve these branches independently. There may be different variants of working with frontal variables holding physical matter or objects rather than information, especially when serving physical movement and possibility of replication or reproduction of their contents in distributed environments.

4.4.4 Nodal variables

Variables of this type, their identifiers starting with capital `N`, are a temporary and exclusive property of the world points visited by SGL scenarios, which can create, change, or remove them.

```
nodal → N{alphameric}
```

Capable of being shared by all scenario branches visiting these nodes, they are created by first assignment to them and stay in the node until removed explicitly or the whole scenario remains active. These variables also cease to exist when nodes they associate with are removed by any scenario reaching them.

4.4.5 Environmental variables

These are special variables with reserved names (all in capitals) which allow us to have access to physical, virtual, and execution worlds when they are navigated by SGL scenarios, also to some parameters of the language interpretation system itself.

```
environmental → TYPE | NAME | CONTENT | ADDRESS | QUALITIES
| WHERE | BACK | PREVIOUS | PREDECESSOR | DOER | RESOURCES |
LINK | DIRECTION | WHEN | TIME | STATE | VALUE | IDENTITY |
IN | OUT | STATUS
```

A brief explanation of their sense and usage is following.

`TYPE`—indicates the type of a node the current scenario step associates with and returns a verbal expression of the node's type (i.e., `virtual`, `physical`, `executive`, or their combination). It can also change the existing node's type by assigning to it another value (simple or combined).

`NAME`—returns name of the current node as a string of characters (only if the node has virtual or executive dimension or both). Assigning to this variable when staying in the node can change the node's name.

`CONTENT`— returns content of the current node (if it has virtual or executive dimension or both) as arbitrary constant (say, any text in quotes, vector or nested structure of multiple texts, etc.) if this content had been assigned to this node previously, when staying in it. Assigning to this variable when staying in the node can change the node's content. In the case of executive nodes (like human, robot, server, etc.), `CONTENT` may return and change, if allowed, some existing specific data like dossier on a human or technical characteristics of a robot.

`ADDRESS`—returns a unique address of the current virtual node (or the one having virtual dimension). This is read-only variable as node addresses are set up automatically by the underlying distributed SGL interpretation system during node's creation, or by an external system (for example, like internet address of the node). The returned address can be remembered and used afterwards for direct hops to this node from any positions of the distributed virtual world, if such hops are allowed for implementation.

QUALITIES—identifies a list of selected formalized physical parameters associated with the current physical position, or node, depending on the chosen implementation and application (for example, these may be temperature, humidity, air pressure, visibility, radiation, noise or pollution level, density, salinity, etc.). These parameters (generally as a list of values) can be obtained by reading the variable. They may also be attempted to be changed (depending on their nature and implementation system capabilities) by assigning new values to QUALITIES, thus locally influencing the world from its particular point.

WHERE—keeps world coordinates of the current physical node (or the one having physical dimension too) in the chosen coordinate system. These coordinates can be obtained by reading this variable. Assigning a new value to this variable (with possible speed added) can cause physical movement into the new position (with preserving node's identity, virtual and/or executive features if any, all its information surrounding, and control and data links with other nodes).

BACK—keeps internal system link to the preceding world node (virtual, executive, or combined one) allowing the scenario to most efficiently return to the previously occupied node, if needed. This variable refers to internal interpretation mechanisms only (its content cannot be lifted, recorded, or changed from the scenario level), and can be used in direct hop operations only.

PREVIOUS— refers to the absolute and unique address of the previous virtual node (or combined one with executive and/or physical dimensions), allowing us to return to the node directly. This return may be on a higher level and therefore more expensive than using BACK, but the content of PREVIOUS, unlike BACK, can be lifted, recorded, and used elsewhere in the scenario (but not changed, similar to ADDRESS).

PREDECESSOR— refers to the name of preceding world node (the one with virtual or executive dimension, visited just before the current one). Its content can be lifted, recorded, and subsequently used, for organization of direct hops to this node too (on highest and most expensive level, however). Assigning to PREDECESSOR in the current node can change the name of the previous node.

DOER—keeps the name of the device (say, laptop, robot, satellite, smart sensor, or a specially equipped human) which interprets the current SGL code in the current world position. This device can be initially chosen for the scenario automatically from the list of recommended devices or just picked up from those expected available. It can also be appointed explicitly by assigning its name to DOER, causing the remaining SGL code (along with its current information surrounding) to move immediately into this device and execute there. (The change of the device can also be done automatically by the distributed SGL interpreter, say, depending on unpredictable circumstances or by dynamic space-conquering optimization.)

RESOURCES—may keep a list of available or recommended resources (human, robotic, electronic, mechanical, etc., by their types or names) which can be used for planning and execution of the current and subsequent parts of the SGL scenario. This list can also contain potential doers which may appear (by their names) in variables `DOER`. The contents of `RESOURCES` can be changed by assignment to it, and in case of automatic distributed SGL interpretation and spatial branching may be replicated or partitioned (or both).

LINK—keeps the name (same as content) of the virtual link which has just been passed. Assigning a new value to it can change the link's content/name. Assigning `nil` or empty to `LINK` removes the link passed.

DIRECTION—keeps direction (along, against, or neutral) of the passed virtual link. Assigning to this variable values like `plus`, `minus`, or `nil` (same as `+`, `-`, or empty) can change its orientation or make the link nonoriented.

WHEN—assigning value to this variable sets up an absolute starting time for the following scenario branch (i.e., starting with the next operation), thus allowing us to suspend and schedule operations and their groups in time.

TIME—returns the current absolute system time as the read-only global variable.

STATE—can be used for explicit setting resultant control state of the current scenario step by assigning to it one of the following constants: `thru`, `done`, `fail`, or `fatal`, which will influence further scenario development from the current world point (and in a broader scale in the case of `fatal`). These control states are also generated implicitly and automatically on the results of success or failure of different operations (belonging to the internal interpretation mechanisms of SGL scenarios). Reading `STATE` will always return `thru` as this could be possible only if the previous operation terminated with `thru` too, thus letting this operation to proceed. A certain state explicitly set up in this variable can also be used at higher levels (possibly, together with termination states of other branches) within distributed control provided by nested SGL rules, whereas assigning `fatal` to `STATE` may cause abortion of multiple distributed processes with associated data.

VALUE—when accessed, returns the resultant value of the latest, i.e., preceding, operation (say, an assignment to it or any other variable, unassigned result of arithmetic or string operation, or just naming a variable or constant). Such explicit or implicit assignment to `VALUE` always leaves its content available to the next operation, which may happen to be convenient by combining different operations traditionally grouped in expressions, within their sequences.

IDENTITY—keeps identity, or color, of the current SGL scenario or its branch, which propagates together with the scenario and influences grouping of different nodal variables under this identity at world nodes reached. This allows different scenarios or their branches with personal identities to be protected from influencing each other, even if they are using similarly named nodal variables in the same world nodes. However, scenarios with different identities can penetrate into each other information fields if they know the other's colors, by temporarily assigning the needed new color to `IDENTITY` at proper stages and world points (say, to perform cooperative or stealth operations) while restoring the previous color afterwards, if needed. Any numerical or string value can be explicitly assigned to `IDENTITY`. By default, different scenarios may be keeping the same value in `IDENTITY` assigned automatically at the start (which may be any, including empty), thus being capable of sharing all information at navigated nodes, unless they change their personal color themselves.

`IN`—special variable requesting and reading data from the outside world in its current point. The received data is becoming the resultant value of the reading operation.

`OUT`—special variable allowing us to issue information from the scenario in its current point to the outside world, by assigning the output value to this variable.

`STATUS`—retrieving or setting the status of (especially doer) node in which the scenario is currently staying (like `engaged` or `vacant`, possibly, with a numerical estimate of the level of engagement or vacancy). This feedback from implementation layer on the SGL scenario layer can be useful for a higher-level supervision, planning and distribution of resources executing the scenario rather than doing this implicitly and automatically.

Other environmental variables for extended applications can be introduced and identified by unique words in all capitals too, or they may use any names if explicitly defined by using special rule, as shown later. As can be seen, most environmental variables are behaving as stationary ones, except `RESOURCES` and `IDENTITY`, which are mobile in nature. The global variable `TIME` may be considered as stationary too, but can also be implemented in the form of individual `TIME` clocks regularly updating their system time copies and propagating with scenarios as frontal variables.

4.5 RULES

The main SGL constructs, called *rules*, are as follows:

```
rule → type | usage | movement | creation | echoing |
verification | assignment | advancement | branching |
transference | exchange | timing | qualifying | grasp
```

The concept of *rule* is dominant in SGL not only for diverse activities on data, knowledge, and physical matter but also for overall management and control of any SGL scenarios. This provides an integral and unified capability for expressing everything that might take place or even come to mind in large dynamic spaces, worlds, and systems, and generally in holistic, highly parallel, and distributed mode. This section describes the main set of SGL rules with summaries of their features.

4.5.1 Type

These rules explicitly assign types to different constructs, with their existing repertoire following.

```
type → global | heritable | frontal | nodal | environmental
| matter | number | string | scenario | constant
```

global, **heritable**, **frontal**, **nodal**, **environmental**—allow different types of variables to have any alphanumeric names, rather than those oriented to self-identification, as explained before. These names will represent variables with needed types in the subsequent scenario developments unless redefined by these rules too.

matter, **number**, **string**, **scenario**, **constant**—allow arbitrary results obtained by the embraced scenario to properly represent the needed values rather than using self-identifiable representations mentioned before.

4.5.2 Usage

These rules explain how to use the information units they embrace, with main variants as follows:

```
usage → address | coordinate | content | index | time |
speed | name | center | range | doer | node | link
```

They are adding certain flexibility to representation of SGL scenarios where strict order of operands in different rules and also presence of them all may not be absolute.

address—identifies the embraced value (which may also be an arbitrary scenario producing this value or values if multiple) as an address of a virtual node.

coordinate—identifies the embraced value as physical coordinates (say, one, two or three-dimensional), which may also be a list of coordinates of a number of physical locations.

content—identifies the embraced operand as a content (or contents) which may, for example, relate to certain values in a list for its search by contents.
index— identifies the embraced operand as an index (or indices), which may represent orders of elements in a list for its search by the index operation.
time—informs that the embraced operand represents time value.
speed—informs that the embraced operand represents a value of speed.
name—identifies the embraced operand as a name (say, of a virtual or executive node or nodes).
center—depending on applications, indicates that virtual address or physical coordinates embraced may relate to the center of some region.
range—identifies virtual or physical distance that can, for example, be used as a threshold for certain operations in distributed spaces, especially those evolving from a chosen or expected center.
doer—identifies the embraced name or any other value as belonging to executive node (like human, robot, server, satellite, smart phone, etc.).
node—(or **nodes**, if more appropriate) identifies the embraced value or values as keeping names of nodes having virtual or/and executive dimensions.
link—(or **links**, if more appropriate) informs that the embraced value or values represent names of links connecting nodes with virtual or/and executive dimensions.

4.5.3 Movement

The movement rules have the following options:

```
movement → hop | hopfirst | hopforth | move | shift |
follow
```

They may result in virtual hopping to the existing nodes (the ones having virtual or/and executive dimensions) or in real movement to new physical locations, subsequently starting the remaining scenario if any (with current frontal variables and control) in the nodes reached. The resultant values of such movements are represented by names of reached nodes (in case of virtual, executive, or combined nodes) or `nil` in case of purely physical nodes, with control state `thru` in them if the movement was successful. If no destinations have been reached, the movement results with state `fail` and value `nil` in the rule's starting node. These rules have the following options.

hop—sets electronic propagation to a node (or nodes) in virtual, execution, or combined spaces (the latter may have physical dimension too), directly or via semantic links connecting them with the starting node. In case of a direct hop, except destination node name or address, special modifier `direct` may be included into parameters of the rule. If the hop is to take place from a node to a particular node via existing link, both destination node name/address and link name (with orientation if appropriate) should be parameters of the rule. This rule can also cause independent and parallel propagation to a number of nodes if there are more than one node connected to the current one by same named links, and only link name mentioned (or given by indicator `all`, for all links involved). In a more general case, parallel hops can be organized from the current node if the rule's parameters are given by a list of possible names/addresses of destination nodes and a list of names of links which may lead to them (`direct` and/or `all` indicators can be used here too). The hop rule may have additional modifiers setting certain conditions for this operation, like `firstcome`, which is based on internal language interpretation mechanism properly marking the visited nodes (for example, to be used for blocking unexpected cycles in network propagations). Another modifiers may link the virtual propagation with some physical parameters of possible combined destination nodes, say, by giving threshold distances to them from the current node (if with physical parameters too).

`hopfirst`—modification of the hop rule allowing it to come to a node only first time (for the scenario with certain identity), which is based on internal interpretation mechanism marking the nodes visited. The use of this rule can be similar to the previous rule hop with modifier `firstcome` (which can also be used in other cases, like new linking to the existing nodes, as mentioned later).

`hopforth`—modification of the previous rule allowing it to hop to a node which is not the one just visited before, i.e., excluding the return to the previous node. It may be considered as a restricted variant of `hopfirst` rule. Both rules can be useful for effective blocking of looping in networked structures for certain scenarios.

`move`—sets real movement in physical world from the current node with physical dimension (which may be combined with virtual and executive ones) to a particular location given by coordinates in a chosen coordinate system. The destination location becomes a new temporary node with no (`nil`) name, which disappears when the current scenario activities leave it for other nodes. The location reached may, however, become a persistent or even permanent node if virtual dimension is also assigned to it (possibly, virtual name too), after which such combined node can become visible from outside, may keep individual nodal variables, and can be entered and shared by other scenario branches. Speed value for the physical propagation by move may be given as an additional parameter.

`shift`—differs from the move only in that movement in physical world is set by deviations of physical coordinates from the current position rather than by their absolute values.

`follow`—allows us to move in virtual, physical, and combined spaces using already (internally, by distributed SGL interpretation) recorded and saved paths from a starting node to the destinations reached, to enter the latter again from the starting node in a simplified way, as will be explained later.

4.5.4 Creation

These rules have the following options:

```
creation → create | linkup | delete | unlink
```

They create or remove nodes and/or links leading to them during distributed world navigation. After termination of the creation rules, their resultant values will correspond to the names of reached nodes with termination states `thru` in them, and the next scenario steps if any will start from all these nodes. After removal of the destination nodes and/or links leading to them, the resultant world position will be the rule's starting node with the same value as before and control state `thru`. If the creation or removal operation fails, its resultant value will be `nil` and control state `fail` in the node the rule started, thus blocking any further scenario development from this node.

`create`—starting in the current world position, creates either new virtual link-node pairs or new isolated nodes. For the first case, the rule is supplied with names and orientations of new links and names of new nodes these links should lead to, which may be multiple. For the second case, the rule has to use modifier `direct` indicating direct nodes creation. If to use modifier `existing` or `passed` for the link-node creation hinting that such nodes already exist or, moreover, have already been passed by this scenario, only links will be created to them by `create`. Same will take place if nodes are given by their addresses, the latter always indicating their existence. The already mentioned modifier `firstcome`, if used, will not allow entering the nodes more than once by the same colored scenario.

`linkup`—restricts the previous rule by creating only links with proper names from the current node to the already existing nodes given by their names or addresses. Using modifier `passed`, if appropriate, may help us to narrow the search of already existing nodes. Also, the modifier `firstcome`, if used, will not allow entering the nodes more than once by same colored scenario or its branch, thus blocking `linkup` operation for this case.

`delete`—starting from the current node removes links together with nodes they should lead to. Links and nodes to be removed should be either explicitly named or represented by modifiers `any` or `all`. Using modifier `direct` instead of link name together with the node name will allow us to remove such node (or nodes) from the current node directly. In all cases, when a node is deleted, its all links with other nodes will be removed too.

`unlink`—removes only links leading to neighboring nodes where, similar to the previous case, they should be explicitly named or modifiers `any` or `all` should be used instead.

The above mentioned creation rules, depending on implementations, can also be used in a broader sense and scale, as *contexts* embracing arbitrary scenarios and influencing hop operations within their scope. This means that the same scenarios will be capable of operating in the creation and deletion modes too, and not only for navigating the existing networks. These contexts can influence both links and nodes when dealing with the existing networks (or just with empty spaces in which such networks should be created from scratch).

4.5.5 Echoing

This class of rules, oriented to various aspects of data and knowledge processing, contains the following rules which may use local and remote values for different operations:

```
echoing → state | rake | order | unit | sum | count | first
| last | min | max | random | average | element | sortup |
sortdown |
reverse | fromto | add | subtract | multiply | divide |
degree | separate | unite | attach | append | common |
withdraw |
increment | decrement | access | invert | apply | location |
distance
```

The listed rules use terminal world positions reached by the embraced scenario with their control states and associated final values (which may be local or arbitrarily remote) to obtain the resultant state and value *in the location where the rule started*. This location will represent the rule's single terminal point from which the rest of the scenario, if any, can develop further. The usual resultant control state for these rules is `thru` (state `fail` occurs only if certain terminal values happen to be unavailable or the result is unachievable, say, like division by zero). Depending on the rule's semantics, the resultant value may happen to be compound like a list of values, which may also be hierarchically nested.

The semantics of different echoing rules is as follows.

`state`—returns the resultant generalized state of the embraced SGL scenario upon its completion, whatever its complexity and space coverage may be. This state being the result of ascending fringe-to-root generalization of terminal states of the scenario embraced, where states with higher power (their sequence from maximum to minimum values as: `fatal`, `thru`, `done`, `fail`) dominate in this potentially distributed and parallel process. The resultant state returned is treated as the *resultant value* on the rule, the latter always terminating with own final control state `thru`, even in the case of resultant `fatal` (thus blocking the spreading destructive influence of `fatal` at the rule's starting point).

rake—returns a list of final values of the scenario embraced in an arbitrary order, which may, for example, be influenced by the order of completion of branches and times of reaching their final destinations. Additionally using unique as modifier, described before, the rule will result in collecting only unique values, i.e., with possible duplicates omitted/removed.

order—returns an ordered list of final values of the scenario embraced corresponding to the order of launching related branches rather than the order of their completion. For potentially parallel branches, these orders may, for example, relate to how they were activated, possibly, with the use of time stamps upon invocation. Similar to the previous rule, modifier unique can be used too for avoiding duplicate values.

unit—returns a list of values while arranging it as an integral parenthesized unit which should not be mixed with elements returned from other branches which may represent integral units too, to form (potentially hierarchical and nested) lists of lists of the obtained values at higher levels. This rule can be combined with rules rake or order to explicitly set up the expected order of returned values in the unit formed. Without unit, at any scenario level, the returned values from different subordinate branches will represent same level mixture of all obtained results.

sum—returns the sum of all final values of the scenario embraced (modifier unique can be used here for summing only unique final values).

count—returns the number of all resultant values associated with the scenario embraced, rather than the values themselves as by the previous rules (modifier unique can be used too for counting only unique values).

first, **last**, **min**, **max**, **random**, **average**— return, correspondingly, the first, last, minimum, maximum, randomly chosen, or average value from all terminal values returned by the scenario embraced. The rules first and last may also need initial ordering of the returned results by previously using or integrating with rule order discussed before, also sortup and sortdown explained below. The modifier unique can be used for all mentioned rules too.

sortup, **sortdown**—return an ordered list of values produced by the embraced scenario operand, starting from minimum or maximum value and ending, correspondingly, with maximum or minimum one.

reverse—changes to the opposite the order of values from the embraced operand.

element— returns the value of an element of the list on its left operand requested by its index (on default) or content (clarifying this by rule content) given by the right operand. If the right operand is itself a list of indices or contents, the result will be a list of corresponding values from the left operand. If element is used within the left operand of assignment, instead of returning values it will be providing an access to them, in order to be updated, as explained later. Each given index representing unique order can return from the left operand one or none value (the latter if the index exceeds total number of elements), whereas each content in the right operand can return from the left operand none (or `nil`), one, or more elements as a list, as there can be repeating values at the left.

`position`— returns the index (or indices) of the list on its left operand requested by the content given by the right operand. There may be more than a single index returned if same content repeats in the list, or if the right operand is itself a list of contents, then each one will participate in the search. The total absence of searched contents in the left operand will result in `nil` value on this rule.

`fromto`— returns an ordered list of digital values by naming its first (operand 1) and last (operand 2) elements as well as step value (operand 3) allowing the next element to be obtained from the previous one. Another modification (depending on implementation) may take into account the starting element, step value, and the number of needed elements in the list.

`add`, **`subtract`**, **`multiply`**, **`divide`**, **`degree`**— perform corresponding operations on two or more operands embraced, each potentially represented by arbitrary scenario with local or remote results. If the operands themselves provide multiple values, as lists, these operations are performed between peer elements of these lists, with the resultant value being multiple, as a list too.

`separate`—separates the left operand string value by the string at the right operand used as a delimiter (in case to be present at the left) in a repeated manner for the left string, with the result being the list of separated substring values. If the right operand is a list of delimiters, its elements will be used sequentially, one after the other, and cyclically unless the string at the left is fully processed/partitioned. If the left operand represents a list of strings, each one is processed by the right operand as above, with the resultant lists of separated values merging into a common list in the order they were produced.

`unite`—integrates the list of values at the left (as strings, or to be converted into strings automatically) by a repeated delimiter as a string too (or a cyclically used list of them) at the right into a united string.

`attach`—produces the resultant string by connecting the right string operand directly to the end of the left one. If operands are lists with more than one element, the attachment is made between their peer elements, receiving the resultant list of united strings. This rule can also operate with more than two operands.

`append`—forms the resultant list from left and right operands by appending the latter to the end of the former as individual elements, where both operands may be lists themselves. More than two operands can be used too for this operation.

`common`— returns intersection of two or more lists as operands, with the result including only same elements of all lists, if any, otherwise ending with `nil`.

withdraw—its returned result will be the first element of the list provided by the embraced operand, which can usually be a variable, along with withdrawing this element from the head of the list (thus simultaneously changing the content of the variable). This rule can have another operand providing the number of elements to be withdrawn in one step and represented as the result. When the embraced list is empty or has fewer elements than needed to be withdrawn, the rule returns `nil` value and terminates with `fail` state.

increment—adds 1 (one) to the value of the embraced operand which will be the result on this rule, thus simultaneously changing the content of the operand itself (this makes sense only if it is a variable, which will be having now the increased value). If another value, not 1 is to be added, the second operand can be employed for keeping this value.

decrement—behaves similar to the previous rule `increment` but subtracts rather than adds 1 from the value of the embraced operand, with the content of the latter simultaneously changed too. Second operand can be used too if the value to be subtracted not equals 1. In all cases if the decrementing result appears to be less than zero, the rule will terminate with `fail` and value `nil`.

access—embracing a scenario or its branch returns a reference to the internal history-based optimized and recorded structure (which may be spatially distributed) leading from the rule-activation node to the reached terminal nodes on the considered scenario. This reference can be remembered (say, in a variable) and subsequently used from the same starting node to reach exactly the same terminal nodes again in an economic and speedy manner. The terminal nodes reentry can be performed by the rule `follow` described before, with its operand reflecting the remembered access reference acquired by `access`.

invert—changes the sign of a value or orientation of a link to the opposite, while producing no effect on zero values or nonoriented links.

apply—organizes application of the first operand as one or a set of rules described above operating jointly from the same starting point (names of which can also be obtained by arbitrary scenario standing for this operand and not only given explicitly) to the same second scenario operand, which may be arbitrary too. If multiple application rules are engaged on the first operand, the obtained results on the second operand can happen to be multiple too.

location–returns world locations of the final nodes reached by the embraced scenario, which mean for virtual nodes their network addresses, and for physical nodes physical coordinates. This may be equivalent to using in the final world positions environmental variables `ADDRESS` or `WHERE` for providing respected open values, with their subsequent collection by other echo rules (directly using `location` may, however, happen to be more convenient in certain cases).

distance–returns distance between two physical points defined by absolute physical coordinates expressed by its parameters, where each one can be represented by an arbitrary scenario.

4.5.6 Verification

This class of rules has the following main variants.

```
verification → equal | nonequal | less | lessorequal |
more | moreorequal | bigger | smaller | heavier | lighter |
longer | shorter | empty | nonempty | belong | notbelong |
intersect | notintersect | yes | no
```

These rules provide control state `thru` or `fail` reflecting the result of concrete verification procedure, also `nil` as own resultant value, while remaining after completion in the same world positions where they started.

equal, **nonequal**, **less**, **lessorequal**, **more**, **moreorequal**, **bigger**, **smaller**, **heavier**, **lighter**, **longer**, **shorter**—make corresponding comparison between left and right operands, which can represent (or result in, if being arbitrary scenarios) information or physical matter/objects, or both. In case of vector operands, state `thru` appears only if all peer values satisfy the condition set up by the rule (except `nonequal`, for which even a single noncorrespondence between any peers will result in overall `thru`). The list of such rules can be easily extended for more specific applications, if supported properly on the implementation level.

empty, **nonempty**—checks for emptiness (i.e., nonexistence of anything, same as `nil`) or nonemptiness (existence) of the resultant value obtained from the embraced scenario.

belong, **notbelong**—verifies whether the left operand value (single or a list with all its elements) belongs as a whole to the right operand generally represented as a list (which may have a single element too).

intersect, **notintersect**—verifies whether there are common elements (values) in the left and right operands, considered generally as lists. More than two operands can be used for these rules too, with at least a single same element to be present in all of them to result `thru` for `intersect`, or no such elements for `notintersect`.

yes—verifies generalized state of the embraced scenario providing own control state `thru` in case of `thru` or `done` from the entire scenario, and control state `fail` in case of resultant `fail` or `fatal` (thus allowing to continue from the node where the rule started only in case of success of the embraced scenario, otherwise terminating).

no—verifies generalized state of the embraced scenario resulting with own control state `thru` in case of `fail` or `fatal` from the scenario, and control state `fail` in case of `thru` or `done` (i.e., allowing to continue from the rule's starting node only in case of failure of the embraced scenario, otherwise terminating).

4.5.7 Assignment

There are two rules of this class:

`assignment` → `assign` | `assignpeers`

These rules assign the result of the right scenario operand (which may be arbitrarily remote, also represent a list of values which can be nested) to the variable or set of variables directly named or reached by the left scenario operand, which may be remote too. The left operand can also provide pointers to certain elements of the reached variables which should be changed by the assignment rather than the whole contents of variables (see also rule `element` mentioned before). These rules will leave control in the same world position they've started, its resultant state `thru` if assignment was successful otherwise `fail`, and the same value (which may be a list) as assigned to the left operand. There are two options of the assignment, as follows.

`assign`—assigns the same value of the right operand (which may be a list of values) to all values (like, say, node names) or variables accessed by the left operand (or their particular elements pointed, which may themselves become lists after assignment, thus extending the lists of contents of these variables). If the right operand is represented by `nil` or empty, the left operand nodes or variables as a whole (or only their certain elements pointed) will be removed.

`assignpeers`—assigns values of different elements of the list on the right operand to different values or variables (or their pointed elements) associated with the destinations reached on the left operand, in a peer-to-peer mode.

4.5.8 Advancement

This class of rules has the following variants:

`advancement` → `advance` | `slide` | `repeat` | `align` | `fringe`

These rules can organize forward or "in depth" advancement in space and time of the embraced scenarios separated by comma. They can evolve within their sequence in synchronous or asynchronous mode using modifiers `synchronous` or `asynchronous` (the second one is optional, as asynchronous is a default mode).

`advance`—organizes stepwise scenarios advancement in physical, virtual, executive, or combined spaces, also in a pure computational space (the latter when staying in the same world nodes with certain data processing, thus moving in time only). For this, the embraced SGL scenario-operands are used in a sequence, as written, where each new scenario shifts to and applies from all terminal world points reached by the previous scenario. The resultant world positions and values on the whole rule are associated with the final steps of the last scenario on the rule (more correctly: of the invocation of all this scenario copies which may operate in parallel by starting from possible multiple points reached by the previous scenario). And the rule's resultant state is a generalization of control states associated with these final steps. If no final steps occur with states `thru` or `done`, the whole advancement on this rule is considered as failed (i.e., with generalized state `fail`), thus resulting without possibility to continue scenario evolution in this direction. On default or with modifier `asynchronous`, the sequence of scenarios on `advance` develops in space and time independently in different directions, with the next scenario from their sequence replicating and starting immediately in all points reached by the previous scenario. This means that different operand scenarios in their sequence may happen to be active simultaneously at the same time, as being developed independently and in parallel, with different times of their completion. With the use of `synchronous` modifier, all invocations of every new scenario (in general: all its multiple copies) in their sequence can start only *after full completion* of all invocations of the previous scenario.

`slide`—works similar to the previous rule unless a scenario in their sequence fails to produce resultant state `thru` or `done` from some world node. In this case the next scenario from the sequence will be applied from the same starting position of the previous failed scenario and so on. The resultant world nodes and values in them will be from the last successfully applied scenarios (not necessarily the same from their sequence, as independently developing in different directions). The results on the whole rule, in their extreme, may even happen to correspond only to the existing value of the node in which the whole rule started (including the node's world position), with state `thru` always being the resultant state in any cases. Both synchronous and asynchronous modes of parallel interpretation of this rule, similar to the previous rule `advance`, are possible, where in the synchronous option, different scenarios (not necessarily their same copies) can simultaneously start only after full completion of the previous parallel steps (also potentially involving different scenarios).

`repeat`—invokes the embraced scenario as many times as possible, with each new iterations taking place in parallel from all final positions with state `thru` reached by the previous invocations. If some scenario iteration fails, its current starting position with its value will be included into the set of final positions and values on the whole rule (this set may have starting positions from different failed iterations, which developed independently in a distributed space). Similar to the previous rule `slide`, in the extreme case, the final set of positions on the whole rule may happen to contain only the position from which the rule started, with state `thru` and value it had at the beginning. By supplying additional numeric modifier to this rule, it is possible to explicitly limit the number of allowed scenario repetitions. Of course, the operand-scenario can be easily internally organized to properly control the allowed number of iterations itself, but using this additional modifier may be useful in some cases. Both synchronous and asynchronous modes of parallel interpretation of this rule similar to the previous rules `advance` and `slide` are possible. In the synchronous mode, at any moment of time only the same scenario iteration (possibly its many copies from different nodes) can develop (whereas some previous ones may have already stopped in other directions). In the asynchronous case, there may be different iterations working in parallel.

`align`—is based on confirmation of full termination of all activities of the embraced operand-scenario in all its final nodes. Only after this, the remaining scenario part, if any, will be allowed to continue from all the nodes reached.

`fringe`—allows us to establish certain constraints (say, by additional parameters) on the terminal world nodes reached by the embraced scenario with final values in them, to be considered as starting positions for the following scenario parts. For example, by comparing values in all terminal nodes and allowing the scenario to continue from a node with maximum or minimum value, integrating this rule with previously mentioned rules `max` or `min` like **`max_fringe`** or **`min_fringe`** can also be possible. Without additional conditions or constrains, this rule is equivalent to the previous one `align`.

For the advancement rules, frontal variables propagate on the forefronts together with advancement of control and operations in distributed spaces, with next scenarios or their iterations picking up frontal variables brought to their starting points by the previous scenarios (or their previous iterations), being also replicated if this control automatically splits into different branches. And the capability and variants of explicit naming and splitting into separate branches will be considered in detail in the next section.

4.5.9 Branching

The rules from this class are as follows:

```
branching → branch | sequence | parallel | if | or | and |
choose | quickest | cycle | loop | sling | whirl | split |
or_sequence | or_parallel | replicate
```

These rules allow the embraced set of scenario operands to develop "in breadth," each from the same starting position, with the resultant set of positions and order of their appearance depending on the logic of a concrete branching rule. The rest of the SGL scenario will be developing from all or some of the positions and nodes reached on the rule. The branching may be static and explicit if we have a clear set of individual operand scenarios separated by comma. It can also be implicit and dynamic, as explained later.

For all branching rules that follow, the frontal variables associated with the rule's starting position will be replicated together with their contents and used independently within different branches, to be inherited by the following scenario, if any, beyond the branching rules. Details of this replication for frontal variables with physical matter rather than information can depend on application and implementation details.

A brief explanation of how these rules work is as follows.

`branch`—the most general and neutral variant of branching, with logical independence of the scenario operands from each other and any possible order of their invocation and development from the starting position (say, ranging from arbitrary to strictly sequential to fully parallel, also any mixture thereof). The resultant set of positions reached with their associated values will unite all terminal positions and values on all scenario operands involved under `branch`. The resultant control state on the whole rule will be based on generalization of the generalized control states on all scenario branches (based on max to min powers of control states: `fatal`, `thru`, `done`, and `fail`, as mentioned before).

`sequence`—organizing strictly sequential invocation of all scenario operands regardless of their success or failure, and launching the next scenario only after full completion of the previous one. The resultant set of positions, values, and rule's global control state will be similar to `branch`. However, the final results may vary due to different invocation order of the scenario operands and possible common information used.

`parallel`—organizing fully parallel development of all scenario operands from the same starting position (at least as much as this can be achieved within the existing environment, resources, and implementation). The resultant set of positions, values, and rule's control state will be similar to the previous two rules, but may not be the same, as explained before.

`if`—may have three scenario operands. If the *first* scenario results with generalized termination state `thru` or `done`, the *second* scenario is activated; otherwise, the *third* one will be launched. The resultant set of positions and associated values will be the same as achieved by the second or third scenarios after their completion. If the third operand-scenario is absent and the first one results with `fail`, or only the first operand is present regardless of its success or failure, the resultant position will be the one the rule started from, with state `thru` and value it had at the start.

`or`—allows *only one* operand scenario with the resulting state `thru` or `done`, without any predetermined order of their invocation, to be registered as resultant, with the final positions and associated values on it to be the resulting ones on the whole rule. The activities of all other scenario operands and all results produced by them will be terminated and canceled. If no branch results with `thru` or `done`, the rule will terminate with `fail` and `nil` value. If used in combination with the previous rules `sequence` and `parallel`, it may have the following features.

or_sequence—will launch the scenario operands in strictly sequential manner, one after the other as they are written, waiting for their full completion before activating the next operand, unless the first one in the sequence replies with generalized state `thru` or `done` (providing the result on the rule as a whole). Invocation of the remaining scenarios in the sequence will be skipped.

or_parallel—activates all scenario operands in parallel from the same current position, with the first one in time replying with generalized `thru` or `done` being registered as the resultant branch for the rule. All other branches will be forcefully terminated without waiting for their completion (or just ignored, depending on implementation, which in general may not be the same due to side effects when working with common resources).

The resultant scenario chosen in all three cases described above with `or` provides its final set of positions with values and states in them as the result on the whole rule. If no scenario operand returns states `thru` or `done`, the whole rule will result with state `fail` in its starting position and `nil` as the resultant value.

and—activates all scenario operands from the same position, without any predetermined order, demanding all of them to return generalized states `thru` or `done`. If at least a single operand returns generalized `fail`, the whole rule results with state `fail` and `nil` value in the starting position while terminating the development of all other branches which may still be in progress. If all operand scenarios succeed, the resulting set of positions unites all resultant positions on all scenario-operands with their associated values. Combining rule `and` with rules `sequence` and `parallel` (as we did for `or`) will clarify their activation and termination order, as follows. (These two options can, in principle, produce dissimilar general results if different scenario operands work with intersecting domains and share information there.)

and_sequence—activates scenario-operands from the same position in the written order, launching next scenario only after the previous one completes with `thru` or `done`, and terminating the whole rule when the current scenario results with `fail`. The remaining scenario operands will be ignored, and all results produced by this and all previous operands will be removed (as far as this can be achievable in a distributed environment).

and_parallel—activates in parallel all scenario operands from the same world position, terminating the rule when the first one in time results with `fail`, while aborting activity of all other operands and removing all results produced by the rule. (The completeness of such cleaning may also depend on its complexity and implementation reality in large distributed spaces, as for the previous case.)

choose—chooses a scenario branch in their sequence *before* its execution, using additional parameters among which, for example, may be its numerical order in the sequence (or a list of such orders to select more than one branch). This rule can also be aggregated with other rules like `first`, `last`, or `random`, by forming combined ones: **`choose_first`**, **`choose_last`**, **`choose_random`**. The resultant set of positions on the rule, their values, and states will be taken from the branch (or branches) chosen.

`quickest`— selects the first branch in time replying its complete termination, regardless of its generalized termination state, which may happen to be `fail` too, even though other branches (to be forcefully terminated now) could respond later with `thru` or `done`. The state, set of positions on this selected branch, and their associated values (if any) will be taken as those for the whole rule. (This rule assumes that different branches are launched independently and in parallel.) It differs fundamentally from the rule `or_parallel` as the latter selects the first in time branch replying with success (i.e., `thru` or `done`) for which, in the worst case, all branches may need to be executed in full to find the branch needed. A modification of `quickest` may have an additional parameter establishing, for example, time limit within which replies are expected or allowed from the branches (and there may be more than one replying branch, which all with `thru` or `done` will be giving integrated result on the rule, otherwise it will terminate with failure). A similar time limit could also be established for the rule `branch` discussed earlier, by considering for the result only branches that reply in proper time. The special timing rules will also be considered later.

`cycle`—repeatedly invokes the embraced scenario from the same starting position until its resultant generalized state remains `thru` or `done`, where on different invocations the same or different sets of resultant positions (with same or different values) may emerge. The resultant set of positions on the whole rule will be an integration of all positions on all successful scenario invocations with their associated values. The following scenario will be developing from all these world positions reached (some or all may be repeating as same starting points) except the ones resulting with state `done`. If no invocation of the embraced scenario succeeds, the resultant state `fail` in the starting position with `nil` value will emerge.

`loop`—differs from the previous rule in that the resultant set of positions on it being only the set produced by the *last* successful invocation of the embraced scenario. (The rule will terminate, as before, with `fail` and `nil` in the starting position if no scenario invocation succeeds.)

`sling`—invokes repeatedly the embraced scenario until it provides state `thru` or `done`, always resulting in the same starting position with state `thru` and its previously associated value when the last iteration results with `fail` (or no invocation was successful at all).

`whirl`—endlessly repeating the embraced scenario from the starting position regardless of its success or failure and ignoring any resultant positions or values produced. External forceful termination of this construct may be needed, like using first in time termination of another, competitive, branch (under the higher-level rule `or_parallel`).

It could also be possible to set an explicit limit on the number of possible repetitions or duration time in the above mentioned cycling-looping-slinging-whirling rules—by supplying them with an additional parameter restricting the repeated scenario invocations, also using the timing rules explained later.

`split`—performs, if needed, additional (and deeper than usual) static or dynamic partitioning of the embraced scenario into different branches, especially in complex and not clear at first sight cases, all starting from the same current position. It may be used alone or in combination with the above mentioned branching rules while preparing separate branches for these rules, ahead of their invocation. Some examples follow.

- If `split` embraces explicit branches separated by commas, it influences nothing as the branches are already declared.
- It the embraced single operand represents broadcasting move or hop (creative or removal including) in multiple directions, the branches are formed from possible variants of its elementary moves or hops, *before* their execution.
- If the rule's operand is an arbitrary scenario (not belonging to the two previous cases), the branches are formed *after* its completion, where each final position reached by the scenario (with its associated values) represents a new branch.
- If the embraced scenario terminates with a single world position but having associated list of values, each value in this list will be treated as an independent position and branch. The rest of SGL scenario will be developing in parallel from each such new branch, with its individual value available by environmental variable `VALUE` described before.

  In a more complex modification, the rule `split` may be applied to the scenario represented as an explicit advancement of different scenarios (one after the other, covered by rule `advance`). In this case, the first such scenario will be split as above, and the remaining ones attached as the following advancement to each obtained branch, thus forming extended branches with the same replicated "tail" (which can be governed altogether by branching rules described above). By the experience with different applications of SGL and its previous variants, such advanced nonlocal splitting mechanism may be useful and effective in different circumstances.

`replicate`—replicates the scenario given by its second operand, providing the number of its copies given by the first operand, with each copy behaving as independent branch starting from the same current world position.

4.5.10 Transference

There are two rules of this class:

```
transference → run | call
```

They organize transference of control in distributed scenarios.

run—transfers control to the SGL code treated as a procedure and being a result of invocation of the embraced scenario (which can be of arbitrary complexity and space coverage, or can just be an explicit constant or variable). The procedure (or a list of them) obtained and activated in such a way can produce a set of world positions with associated values and control states as the result on the rule, similar to other rules. In case of failure to treat and activate results of the embraced operand as an SGL scenario, this rule will terminate with value `nil` and state `fail` in the node it started.

call—transfers control to the code produced by the embraced scenario which may represent activation of external systems (including those working in other formalisms). The resultant world position on `call` will be the same where the rule started, with value in it corresponding to what has been returned from the external call and state `thru` if the call was successful, otherwise `nil` and `fail` of the latter two.

4.5.11 Exchange

exchange → input | output | send | receive | emit | get

input—provides input of external information or physical matter (objects) on the initiative of SGL scenario, resulting in the same position but with value received from the outside. The rule may have an additional argument clarifying a particular external source from which the input should take place. The rule extends possibilities provided by reading from environmental variable `IN` explained before.

output—outputs the resultant value obtained by the embraced scenario, which can be multiple, with the same resultant position as before and associated value sent outside (in case of physical matter, the resultant value may depend on the applications). The rule may have an additional pointer to a particular external sink. The rule extends possibilities provided by assignment to the previously explained environmental variable `OUT`.

send—staying in the current position associated with physical, virtual, executive (or combined) node, sends information or matter obtained by the scenario on the first operand to other similar node given by name, address, or coordinates provided by the second operand, assuming that a companion rule `receive` is already engaged there. The rule may have an additional parameter setting acceptable time delay for the consumption of this data at the receiving end. If the transaction is successful, the resultant position will be the same where the rule started with state `thru` and value sent (in case of physical matter, this may depend on application and implementation capabilities), otherwise `nil` and state `fail`.

receive—a companion to rule send, naming the source of data to be received from (defined similarly to the destination node in send). Additional timing (as a second operand) may be set too, after expiration of which the rule will be considered as failed. In case of successful receipt of the data, the rule will result in the same world position and the value obtained (information or matter) from send and state thru, otherwise will terminate with value nil and state fail.

emit—depending on implementation and technical capabilities can trigger nonlocal to global continuous broadcasting of the data obtained by the embraced scenario, possibly, with tagging of this source (like setting the emission frequency). Another operand providing time allowed for this broadcasting may be present too. No feedback from possible consumers of the sent data is expected. Will terminate in the application node with the broadcast value and state thru in case of success, otherwise with nil and fail.

get—tries to receive data which can be broadcast from some source (say, identified by its tag or frequency), with resultant value as the received data and state thru in the application node, otherwise with nil and state fail. Similar to the previous rule, additional operand can be introduced for limiting the activity time of this rule. No synchronization with the data emitting node is expected.

4.5.12 Timing

The following two options are available for this rule:

```
timing  →  sleep | allowed
```

These rules are dealing with conditions related to a time interval for the scenarios they embrace.

sleep—establishes time delay defined by the embraced scenario operand, with suspending activities of this particular scenario branch in the current node. The rule's starting position and its existing value, also state thru, will be the result on the rule after the time passed. Similar time delay of the related branch can also be achieved by assigning the current absolute time (say, received from environmental variable TIME) incremented by the needed delay value to environmental variable WHEN described before.

allowed—sets time limit by the first operand for an activity of the scenario on the second operand. If the scenario terminates before expiration of this time frame, its resultant positions with values and states will define the result on this rule. Otherwise, the scenario will be forcibly aborted, with state fail and value nil as the rule's result in its starting position.

4.5.13 Qualification

This class containing the following rules:

`qualification → contain | release | trackless | free | blind | quit | abort | stay | lift | seize | exit`

These rules are providing certain qualities or abilities, also setting constraints or restrictions to the scenarios they embrace, as follows.

contain—restricts the spread of abortive consequences caused by control state `fatal` within the ruled scenario. This state may appear automatically and accidentally in different scenario development points or can be assigned explicitly to environmental variable `STATE`, triggering emergent completion of multiple scenario processes and removal of temporary data associated with them. The resultant position on the rule `contain` having state `fatal` inside its scenario will be the one it started from, with value `nil` and state `fail`. Without occurrence of `fatal`, the resultant positions, their values and states on the rule will be exactly the same as from the scenario embraced and normally terminated. The destructive influence from state `fatal` is also automatically stopped if the scenario in which it may appear is covered by rule `state` (converting any embraced control state into a value), also rules `yes` and `no` (first changing the embraced state into `fail` and second into `thru` in case of `fatal`), as described before. But after these three rules the resultant world positions always correspond to the single rule's starting node regardless of what the embraced scenario produces, whereas `contain` without `fatal` results in exactly the same final positions and values (which may be many) as the scenario embraced.

release—allows the embraced scenario to develop free from the main scenario, abandoning bilateral control links with it (the main scenario after the rule's activation "will not see" this construct any more). The released, now independent, scenario will develop in a usual way using its standard subordination and control mechanisms. For the main scenario, this rule will be immediately considered as terminated in the point it started, with state `thru` and original value there.

trackless—allows the whole embraced scenario to develop absolutely free from any previous stages (i.e., without saving any control and information links with them) like a real virus, being in some sense extreme and global variant of `release`. Under this rule, heritable variables, also environmental ones like `BACK`, `PREVIOUS`, `PREDECESSOR`, `LINK`, `DIRECTION` and `VALUE`, will always result with `nil`, but frontal variables carrying information between different stages will work as usual.

free—differs from the previous case in that despite its independence and control freedom from the main scenario, as before, the embraced scenario will nevertheless be obliged to return the final data obtained in its terminal positions to the main scenario (if such a request issued by certain rules covering the part under `free`).

blind, quit, abort—after full completion of the embraced scenario, these rules result in the same position the rule started with respective states done, fail, or fatal, thus preventing further scenario development from this point (also triggering nonlocal termination and cancellation processes in case of fatal). These rules may represent more economic solutions than explicit termination of all final branches of the embraced scenario with states done or fail. If the ruled scenario is omitted (i.e., rule names standing alone), these rules will be equivalent to assigning the related states to environmental variable STATE in the position they've started.

stay—whatever the scenario embraced and its evolution in space, the resultant position will always be the same this rule started from (and not termination positions of the ruled scenario), with value nil and state thru in it. If the ruled scenario is omitted, this rule standing alone just represents an empty operation in the current point or assignment state thru to variable STATE in it. Such empty operation with stay can also be used for declaring a new scenario branch which can be independently followed by the rest of the scenario starting from this point.

lift—lifts blocking of the further scenario developments set up by states done in the embraced scenario wherever it happened to emerge (including equivalent effect caused by rules blind), substituting them with thru and allowing further developments from all such positions by the rest of the scenario, which may be massive and space-distributed.

seize—establishes, or seizes, an absolute control over the resources associated with the current virtual, physical, executive, or combined node, blocking these from any other accesses and allowing only the embraced scenario to work with them (thus preventing possible competition for the node's assets which may lead to unexpected results). This resource blockage is automatically lifted after the embraced scenario terminates. The resultant set of positions on the rule with their values and states will be the same as from the scenario embraced. If the node has already been blocked by some other scenario exercising its own rule seize, the current scenario will be waiting for the release of the node. If more than two scenarios are competing for the node's resources, they will be organized in a FIFO manner at the node.

exit—terminated the innermost loop in which it is included.

4.5.14 Grasping

The rule's identifier can be expressed not only as a directly given name but also by the result produced by a scenario of any complexity and treated as rule's name. It can also be a compound one, integrated from multiple names provided by different scenarios, so in general we may have the following:

rule → grasp → constant | variable | rule({ grasp, })

Under this extended definition, resulting from recursive SGL syntax, additional parameters can also associate with the rule's names, before embracing the main scenario operands. Such aggregation can simplify the structure of SGL scenarios, also making them more flexible and adjustable to changing goals and environments in which they operate.

4.6 EXAMPLES OF SPATIAL SCENARIOS IN SGL

We will show here solutions, with their analyses, of some practical problems from different areas that can be effectively described and solved in SGL. More details on similar and many other examples using SGL and its previous versions can be found in [1–41].

4.6.1 Network management

Finding any path between nodes `a` and `f` in the network of Figure 4.2, as follows.

```
advance(hop(a),
        repeat(advance(
           hopfirst(links(all)),
           if(equal(NAME, f), done))))
```

Let us express the structure of this scenario with the following generalized operations and rules:

```
g1 → hop(a)
g2 → hopfirst(links(all))
g3 → if(equal(NAME, f), done)
a1, a2 → advance
a1(g1,rp(a2(g2,g3))) → full scenario
```

Its spatiochart is shown in Figure 4.3 with double invocation of rule `advance` under different names `a1` and `a2`.

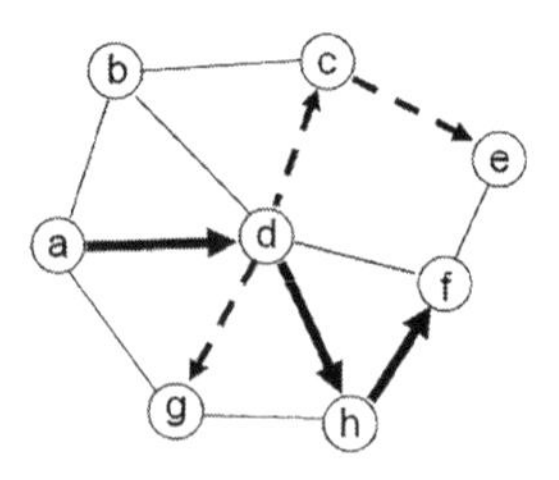

Figure 4.2 Finding a path between two network nodes.

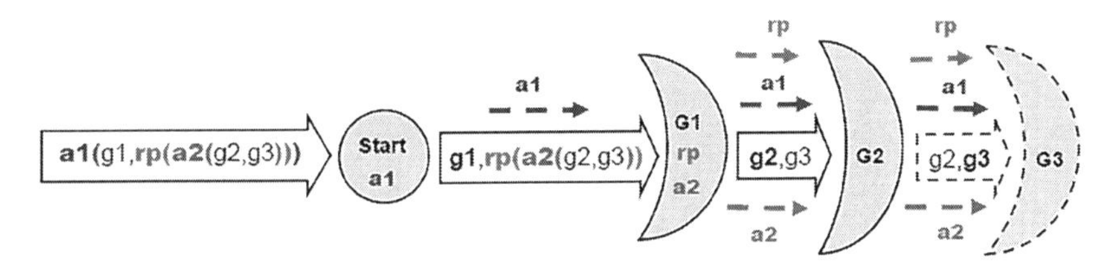

Figure 4.3 Path finding spatiochart.

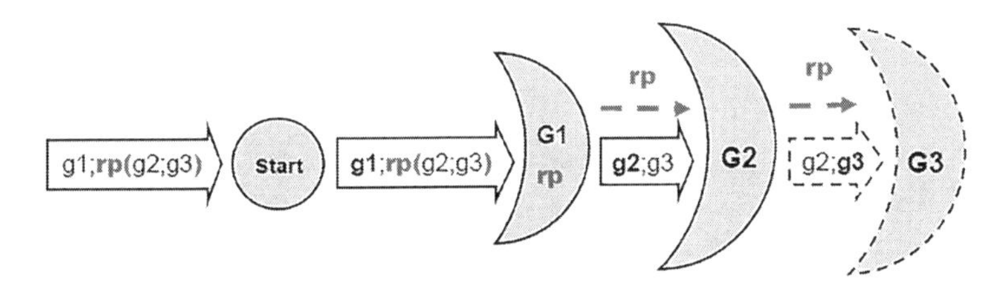

Figure 4.4 Path finding spatiochart simplification.

Simplification where rule `advance` (there is also another and more complex option of advancement like `slide`) can be expressed by just using semicolon, as follows, with the resultant chart shown in Figure 4.4 (where `G2` and `G3` are representing the same network nodes, and `g2` and `g3` will be repeated until node `f` is found).

```
hop(a); repeat(hopfirst(links(all));
              if(NAME == f, done))
```

We can also use spatiochart expressing more details of this scenario, as follows:

```
g1 → hop(a)
g2 → hopfirst(links(all))
g3 →  NAME == f
G4 → done
if → if
g1;rp(g2;if(g3,g4)) → full scenario
```

The resultant spatiochart is in Figure 4.5.

We may decide to collect the passed path and organize its output at the destination node, as follows.

```
hop(a); frontal(Path = NAME);
repeat(hopfirst(links(all)); Path &&= NAME;
       if(NAME == f, blind(output(Path))))
```

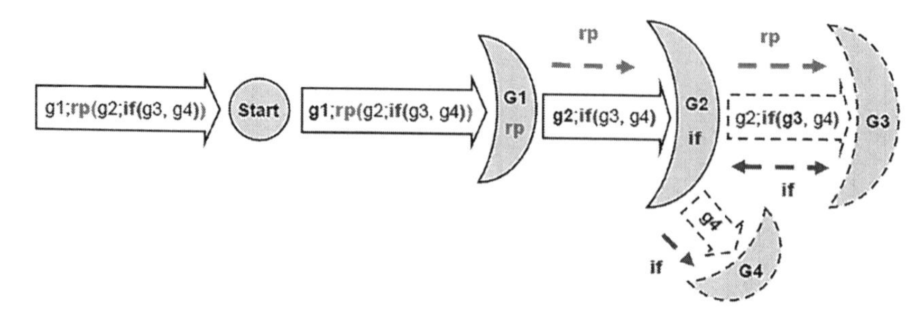

Figure 4.5 More detailed path finding spatiochart.

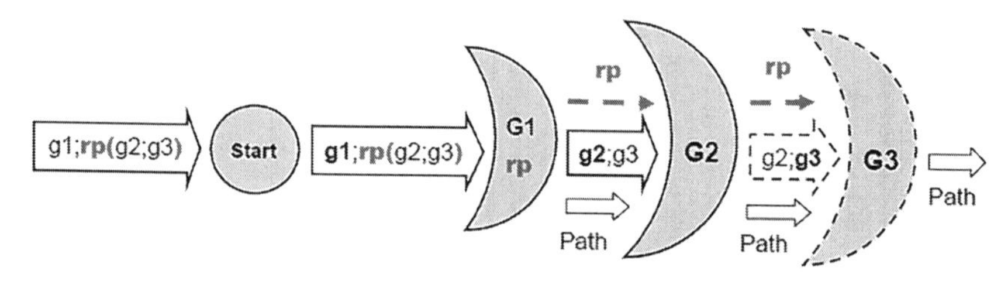

Figure 4.6 Path collecting spatiochart in final node.

using the following details for the chart:

```
g1 → hop(a); frontal(Path = NAME)
g2 → hopfirst(links(all)); Path &&= NAME
g3 → if(NAME == f, blind(output(Path)))
rp → repeat
g1;rp(g2;g3) → full scenario
```

The related spatioachart is shown in Figure 4.6.

Arbitrary path found between these two nodes (may not be optimal like the one in Figure 4.2) is printed in node `f` like: `(a, d, h, f)`. Such path can also be returned and issued in the starting node `a` as:

```
hop(a); frontal(Path = NAME);
output(repeat(hopfirst(link(any)); Path &&= NAME;
                         if(NAME == f, blind(Path))))
```

Details:

```
g1 → hop(a); frontal(Path = NAME)
g2 → hopfirst(links(all)); Path &&= NAME
g3 → if(NAME == f, blind(Path))
out → output
g1;out(rp(g2;g3)) → full scenario
```

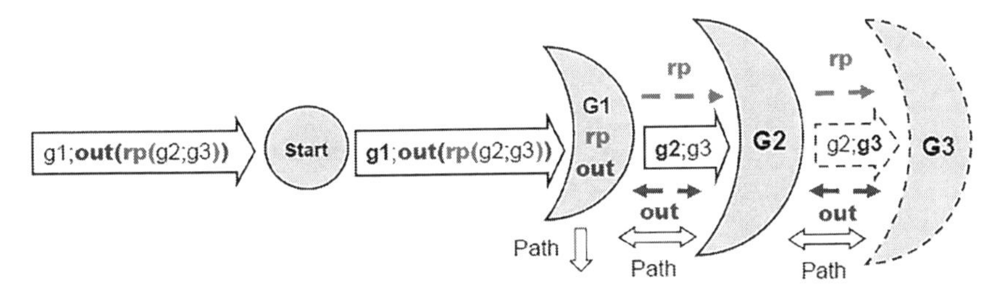

Figure 4.7 Path retuning spatiochart in network starting node.

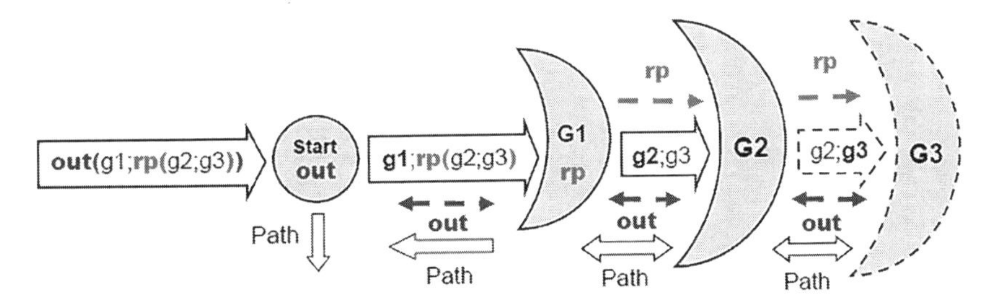

Figure 4.8 Path retuning spatiochart in scenario starting node.

The related chart is in Figure 4.7.

Such path can also be returned and issued at the system entry, as below and in Figure 4.8):

```
output(hop(a); frontal(Path = NAME);
      repeat(hopfirst(link(any)); Path &&= NAME;
                    if(NAME == f, blind(Path))))
```

SG solution analysis

The described solutions on a distributed network are entirely based on its spatial navigation by SGL self-evolving scenarios with finding a path between the nodes needed, also collecting and printing this path in its final or starting nodes. Such integral parallel and fully distributed solutions are much superior to traditional methods of representing distributed computations in the form of multiple parts or agents exchanging messages.

4.6.2 Human-robotic collectives

We can easily organize any collectives from human, robotic, or mixed units operating under spatial scenarios, as symbolically shown in Figure 4.9.

Randomized collective group movement, starting in any node, with minimal `Range` distance allowed between units when moving, can be organized

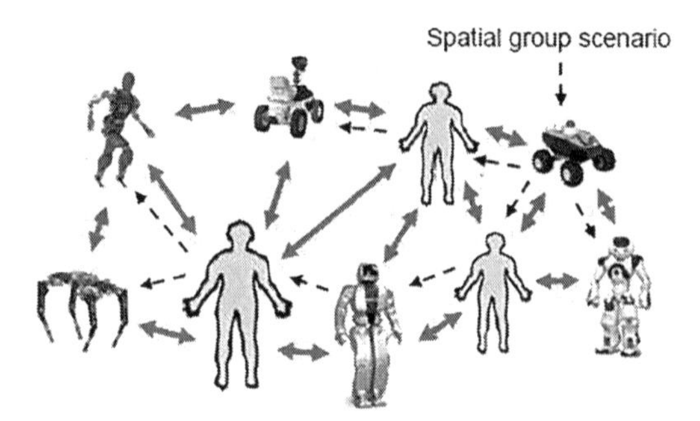

Figure 4.9 Unified human-robotic team.

as follows, where each unit by discovering some dangerous objects or situations issues a corresponding `alarm` message or sound.

```
hop(all);
nodal(Limits = (dx(0,8), dy(-2,5)),
      Range = 100, Shift);
repeat(Shift = random(Limits);
      if(empty(Shift, Range), WHERE += Shift);
      if(analyze(Seen), output(alarm)))
```

Chart-oriented details:

```
g1 → hop(all);
     nodal(Limits = (dx(0,8), dy(-2,5)),
           Range = 100, Shift);
G2 → Shift = random(Limits);
g3 → empty(Shift, Range)
g4 → WHERE += Shift
g5 → analyze(Seen)
g6 → output(alarm)
if1, if2 → if
g1;rp(g2;if1(g3,g4);if2(g5,g6)) → full scenario
```

The related spatiochart is in Figure 4.10.

The following is a more dynamic solution, where scenario starts from any unit, and its gradual spreading to other units is organized in parallel with activity of the units that have been already reached, with the latter now operating fully independently.

```
hop(any_unit);
repeat(
  hop_first(all_neighbors);
  nodal(Limits = (dx(0,8), dy(-2,5)),
        Range = 100, Shift);
  free(repeat(
         Shift = random(Limits);
```

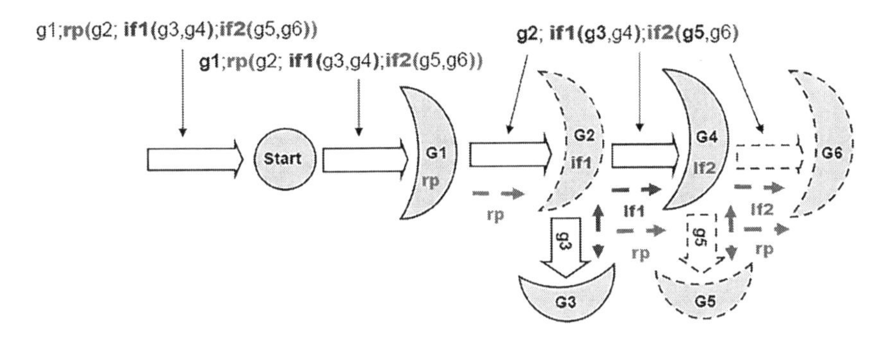

Figure 4.10 Collective behavior starting in all units.

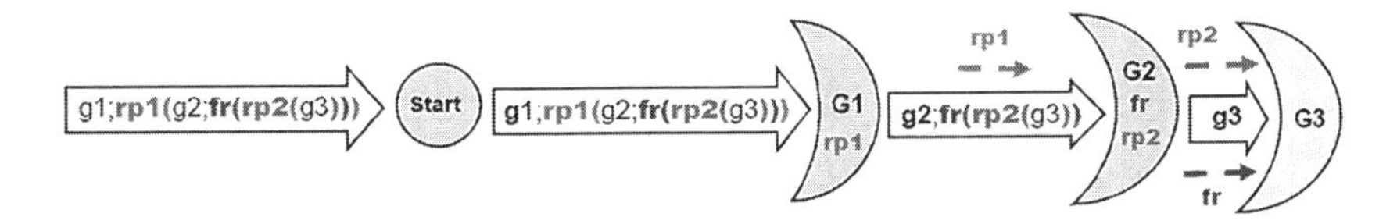

Figure 4.11 Collective behavior gradually covering all units.

```
        if(empty(Shift, Range), WHERE += Shift);
        if(analyze(Seen), output(alarm))))
```

Chart-oriented details:

```
g1   → hop(any_unit)
rp1  → repeat
g2   → hop_first(all_neighbors)
       nodal(Limits = (dx(0,8), dy(-2,5)),
            Range = 100, Shift)
fr   → free
rp2  → repeat
g3   → Shift = random(Limits);
       if(empty(Shift, Range), WHERE += Shift);
        if(analyze(Seen), output(alarm))
g1;rp1(g2;fr(rp2(g3))) → full structure:
```

The related spatiochart is in Figure 4.11.

SG solution analysis

This scenario starts from any human or robotic unit and then covers all reachable units by flooding them in parallel, regardless of their number which may be arbitrarily large and not known in advance. It immediately tasks each reached unit with the collective movement and observation functionality, without waiting for full completion of the scenario distribution. This holistic self-evolving spatial solution is also much superior to traditional parallel computations representing the system as a collection of agents exchanging messages.

4.6.3 Spreading and fighting viruses

We will show here a sketch in SGL of how to model massive spread of COVID-like virus, also distribution and use of antivirus vaccine with its influence on further virus development. In Figure 4.12 the virus is assumed originating from a source `S` and initial vaccination starting from locations `V1` and `V2`, with both processes spreading worldwide in parallel.

The following scenario describes unlimited virus spread in a random mode with certain `Breadth` from any reached point. With this spread being chaotic and capable of returning to the previous areas along with covering new regions, while being halted in the already visited points or where the vaccination took place.

```
move(S);
frontal(Breadth = number1,
        Limits = (Xmin1_Xmax1, Ymin1_Ymax1));
repeat(
  replicate(Breadth, shift(random(Limits)));
  if(STATUS == not(protected), STATUS = infected))
```

Chart-oriented details:

```
g1 → move(S);
     frontal(Breadth = number1,
             Limits = (Xmin1_Xmax1, Ymin1_Ymax1))
rp → repeat
g2 → replicate(Breadth, shift(random(Limits)))
g3 → if(STATUS == not(protected), STATUS = infected)
frontals → Breadth, Limits
g1;rp(g2;g3) → full structure:
```

The corresponding chart shown in Figure 4.13.

The next scenario describes the world coverage by vaccination which, starting from some centers `V1` and `V2`, distributes randomly too, but in a

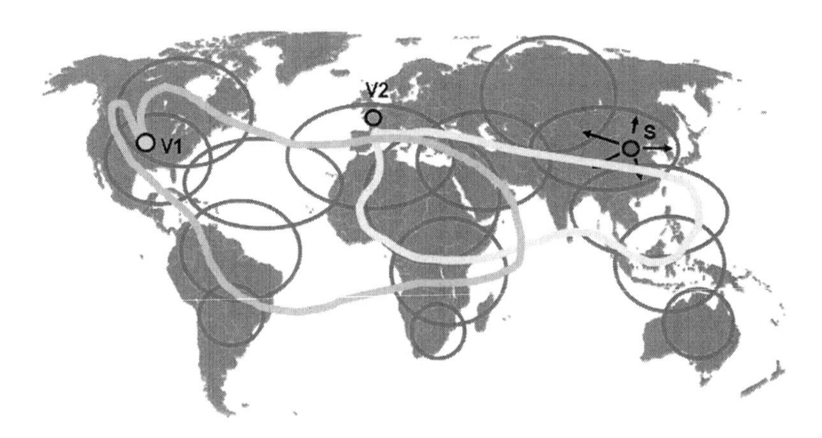

Figure 4.12 Spreading and fighting viruses worldwide.

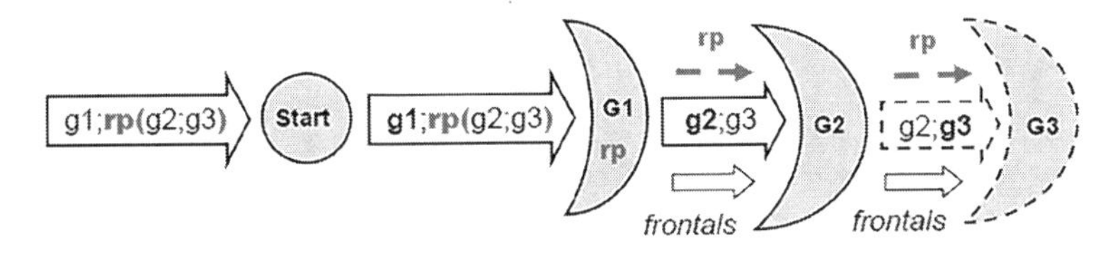

Figure 4.13 Virus spreading chart.

more ordered way, by always extending the vaccinated space, in contrast with the fully chaotic manner by the virus, as before.

```
move(V1, V2);
nodal(Distance, Shift);
frontal(Start = WHERE, Far = 0, Breadth = number2,
        Limits = (Xmin2_Xmax2, Ymin2_Ymax2));
repeat(
  replicate(Breadth,
    repeat(
      Shift = random(Limits);
      Distance = distance(Start, WHERE + Shift);
      Distance < Far));
  shift(Shift); Far = Distance;
  STATUS = protected)
```

Chart-oriented details:

```
g1 → move(V1, V2);
g2 → nodal(Distance, Shift);
     frontal(Start = WHERE, Far = 0,
             Breadth = number2,
             Limits = (Xmin2_Xmax2, Ymin2_Ymax2));
rp → repeat
g3 → replicate(Breadth,
        repeat(
          Shift = random(Limits);
          Distance = distance(Start, WHERE + Shift);
          Distance < Far));
G4 → shift(Shift); Far = Distance;
     STATUS = protected
frontals → Start, Far, Breadth, Limits
g1;g2;rp(g3;g4) → full scenario:
```

The resultant spatiochart is shown in Figure 4.14.

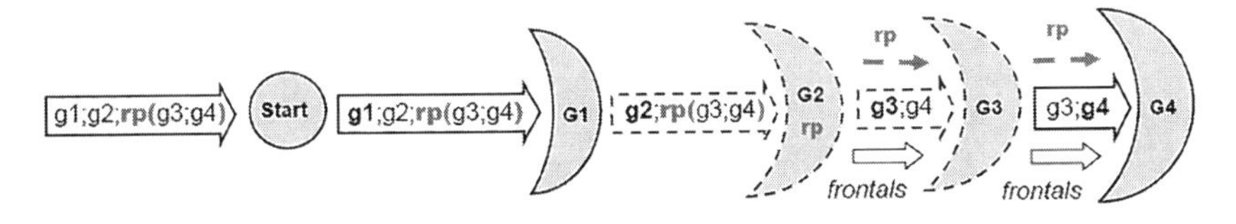

Figure 4.14 Vaccination distribution chart.

SG solution analysis

These are highly dynamic and fully distributed solutions, where the system in which parallel and concurrent virus and vaccination processes are taking place is outlined and formed dynamically in SGL, where fully interpreted SG model is itself developing as an almighty virus which can do everything in fully distributed environments. Such processes may also relate to global warming, weather crises, dynamics of market economies, military conflicts, cultural and religious clashes, spreading different cultures and beliefs, and so on.

Many more on virus and antivirus vaccine simulation and management, and other global self-evolving processes, can be found in many existing publications on SGT and SGL.

4.7 CONCLUSION

We have provided full details of the Spatial Grasp Language, SGL, suitable for parallel processing in large distributed terrestrial and celestial environments both virtual-networked, and physical, which can be arbitrarily large and have no borders. Having only three conceptual components like *constants*, *variables*, and *rules*, and universal recursive syntax, it allows us to describe and organize arbitrary complex processes in a variety of distributed systems. The language allows us to grasp complex problems in different spaces and their solutions on topmost semantic level in pattern-matching mode, while allowing at the same time to describe and deal with any details needed, on all levels. The peculiarity of high-level language constructs and their combinations allow us to shift most of system management routines to the automatic and intelligent interpretation level. The language space-grasping philosophy and organization also allows it to be easily extendable to any other classes of problems by simply adding new specific rules within the same recursive syntax.

REFERENCES

1. P.S. Sapaty, *Mobile Processing in Distributed and Open Environments*, John Wiley & Sons, New York, 1999.
2. P.S. Sapaty, Ruling Distributed Dynamic Worlds, John Wiley & Sons, New York, 2005.
3. P.S. Sapaty, Managing Distributed Dynamic Systems with Spatial Grasp Technology, Springer, 2017.
4. P.S. Sapaty, Holistic Analysis and Management of Distributed Social Systems, Springer, 2018.
5. P.S. Sapaty, Complexity in International Security: A Holistic Spatial Approach, Emerald Publishing, 2019.

6. P.S. Sapaty, Symbiosis of Real and Simulated Worlds under Spatial Grasp Technology, Springer, 2021. 251 p.
7. P.S. Sapaty, A Brief Introduction to the Spatial Grasp Language (SGL), *Journal of Computer Science & Systems Biology* 9(2) (2016).
8. P.S. Sapaty, Spatial Grasp Language for distributed management and control, *Mathematical Machines and Systems* 3 (2016).
9. P.S. Sapaty, Spatial Grasp Language (SGL), *Advances in Image and Video Processing* 4(1) (2016).
10. P.S. Sapaty, Spatial Grasp Language (SGL) for distributed management and control, *Journal of Robotics, Networking and Artificial Life* 4(2) (2016).
11. P.S. Sapaty, M. Sugisaka, A Language for Programming Distributed Multi-Robot Systems, in *Proc. of The Seventh International Symposium on Artificial Life and Robotics (AROB 7th '02)*, January 16–18, 2002, B-Com Plaza, Beppu, Oita, Japan, pp. 586–589.
12. P.S. Sapaty, Mobile programming in WAVE, *Mathematical Machines and Systems* 1 (1998), pp. 3–31.
13. P.S. Sapaty, P.M. Borst, An overview of the WAVE language and system for distributed processing in open networks, Technical Report, Dept. Electronic & Electrical Eng, University of Surrey, June 1994.
14. P.S. Sapaty, A brief introduction to the WAVE language, Report No. 3/93, Faculty of Informatics, University of Karlsruhe, 1993.
15. P.S. Sapaty, WAVE-1: A new ideology of parallel processing on graphs and networks, *Future Generations Computer Systems* Vol. 4, North-Holland, 1988.
16. P.S. Sapaty, The WAVE-1: A new ideology and language of distributed processing on graphs and networks, *Computers and Artificial Intelligence* 5, 1987.
17. P.S. Sapaty, A wave language for parallel processing of semantic networks, *Computers and Artificial Intelligence* 5(4) (1986).
18. P.S. Sapaty, The WAVE-0 language as a framework of navigational structures for knowledge bases using semantic networks, Proc. USSR Academy of Sciences. Technical Cybernetics, No. 5, 1986 (in Russian).
19. P.S. Sapaty, A wave approach to the languages for semantic networks processing, *Proc. Int. Workshop on Knowledge Representation.* Section 1: Artificial Intelligence, Kiev, 1984 (in Russian).
20. P.S. Sapaty, On possibilities of the organization of a direct intercomputer dialogue in ANALYTIC and FORTRAN languages, Publ. No. 74–29, Inst. of Cybernetics Press, Kiev, 1974 (in Russian).
21. P.S. Sapaty, Spatial Grasp Language for distributed management and control, MMC, 3 (2016).
22. P.S. Sapaty, Spatial Grasp Language (SGL), *Advances in Image and Video Processing* 4(1) (2016). http://scholarpublishing.org/index.php/AIVP/
23. P.S. Sapaty, Mosaic warfare: from philosophy to model to solutions, *International Robotics & Automation Journal* 5(5) (2019). https://medcraveonline.com/IRATJ/IRATJ-05-00190.pdf
24. P.S. Sapaty, Advanced terrestrial and celestial missions under spatial grasp technology, *Aeronautics and Aerospace Open Access Journal* 4(3) (2020). https://medcraveonline.com/AAOAJ/AAOAJ-04-00110.pdf
25. P.S. Sapaty, Spatial management of distributed social systems, *Journal of Computer Science Research* 2(3) (2020, July). https://ojs.bilpublishing.com/index.php/jcsr/article/view/2077/pdf

26. P.S. Sapaty, Towards global nanosystems under high-level networking technology, *Acta Scientific Computer Sciences* 2(8) (2020). https://www.actascientific.com/ASCS/pdf/ASCS-02-0051.pdf
27. P.S. Sapaty, Symbiosis of distributed simulation and control under spatial grasp technology, SSRG *International Journal of Mobile Computing and Application (IJMCA)* 7(2) (2020, May–August. http://www.internationaljournalssrg.org/IJMCA/2020/Volume7-Issue2/IJMCA-V7I2P101.pdf
28. P.S. Sapaty, Global network management under spatial grasp paradigm, *International Robotics & Automation Journal* 6(3) (2020). https://medcraveonline.com/IRATJ/IRATJ-06-00212.pdf
29. P.S. Sapaty, Global network management under spatial grasp paradigm, *Global Journal of Researches in Engineering: J General Engineering* 20(5) (2020) Version 1.0. https://globaljournals.org/GJRE_Volume20/6-Global-Network-Management.pdf
30. P.S. Sapaty, Symbiosis of Real and Simulated Worlds Under Global Awareness and Consciousness, abstract at The Science of Consciousness Symposium TSC 2020. https://eagle.sbs.arizona.edu/sc/report_poster_detail.php?abs=3696
31. P.S. Sapaty, Fighting global viruses under spatial grasp technology, *Transactions on Engineering and Computer Science* 1(2) (2020). https://gnoscience.com/uploads/journals/articles/118001716716.pdf
32. P.S. Sapaty, Symbiosis of virtual and physical worlds under spatial grasp technology, *Journal of Computer Science & Systems Biology* 13(6) (2020). https://www.hilarispublisher.com/open-access/symbiosis-of-virtual-and-physical-worlds-under-spatial-grasp-technology.pdf
33. P.S. Sapaty, Space debris removal under spatial grasp technology, *Network and Communication Technologies* 6(1) (2021). https://www.ccsenet.org/journal/index.php/nct/article/view/0/45486
34. P.S. Sapaty, Spatial grasp as a model for space-based control and management systems, *Mathematical Machines and Systems* 1 (2021), 135–138. http://www.immsp.kiev.ua/publications/articles/2021/2021_1/Sapaty_book_1_2021.pdf
35. P.S. Sapaty, Managing multiple satellite architectures by spatial grasp technology, *Mathematical Machines and Systems* 1 (2021), 3–16. http://www.immsp.kiev.ua/publications/eng/2021_1/
36. P.S. Sapaty, Spatial management of large constellations of small satellites, *Mathematical Machines and Systems* 2 (2021). http://www.immsp.kiev.ua/publications/articles/2021/2021_2/02_21_Sapaty.pdf
37. P.S. Sapaty, Global management of space debris removal under spatial grasp technology, *Acta Scientific Computer Sciences* 3(7) (2021, July). https://www.actascientific.com/ASCS/pdf/ASCS-03-0135.pdf
38. P.S. Sapaty, Space debris removal under spatial grasp technology, *Network and Communication Technologies* 6(1) (2021). https://www.ccsenet.org/journal/index.php/nct/article/view/0/45486
39. P.S. Sapaty, Spatial grasp model for management of dynamic distributed systems, *Acta Scientific Computer Sciences* 3(9) (2021). https://www.actascientific.com/ASCS/pdf/ASCS-03-0170.pdf
40. P.S. Sapaty, Spatial grasp model for dynamic distributed systems, *Mathematical Machines and Systems* 3 (2021). http://www.immsp.kiev.ua/publications/articles/2021/2021_3/03_21_Sapaty.pdf
41. P.S. Sapaty, Development of space-based distributed systems under spatial grasp technology, *Mathematical Machines and Systems* 4 (2021).

Chapter 5

Elementary constellation operations under Spatial Grasp Technology (SGT)

5.1 INTRODUCTION

The aim of this chapter is to show how different problems emerging with the launch of large satellite constellations [1–11] can be resolved with the help of high-level model and technology already tested on many distributed systems applications [12–34]. The number of satellites in low Earth orbits is predicted to grow dramatically in the coming years due to the launch of planned satellite constellations, which, despite high attractiveness, raise enormous amount of problems for their effective implementation. The chapter investigates the developed high-level Spatial Grasp Technology (SGT) for its principal applicability for effective organization and management of large satellite constellations, where SGT can convert the whole constellation into a self-organized and goal-oriented system, if needed, with reduced expensive communication with ground antennas and systems.

The rest of the chapter is organized as follows. *Section 5.2* repeats a brief summary of SGT described in detail in Chapters 3 and 4, which will be useful for a quicker understanding of the following sections. *Section 5.3* describes how to integrate satellite constellations into a capable system under SGT. *Section 5.4* shows how to provide broadcasting of executive orders to all satellites from a ground station under SGT. *Section 5.5* shows how supplying satellites with SGL interpreters, which can communicate directly with each other as an integral system and also with the ground stations, may convert the constellation into a self-organized and entirely space-located system, with significant simplification of ground antennas and reduction of their numbers. *Section 5.6* shows some basic operations over satellite constellations in SGL in a virus-like self-spreading parallel mode, which include broadcasting executive orders to all satellites via direct and changeable communications between them, collecting and returning accumulated data by all satellites, and constellation repositioning and restructuring which may be emergent. *Section 5.7* mentions more complex and advanced constellation solutions which can be provided under SGT, and which will be discussed in the subsequent chapters. *Section 5.8* concludes the chapter.

DOI: 10.1201/9781003230090-5

5.2 A BRIEF SUMMARY OF SPATIAL GRASP TECHNOLOGY

Within SGT described in Chapters 3–4, also in more detail in [12–18], a high-level scenario for any task to be performed in a distributed world is represented as an *active self-evolving pattern* rather than a traditional program, sequential or parallel one. This pattern, written in a recursive high-level Spatial Grasp Language (SGL) and expressing top semantics of the problem to be solved, can start from any point of the world. Then it spatially *propagates, replicates, modifies, covers, and matches* the distributed world in a parallel wavelike mode, while echoing the reached control states and data found or obtained for making decisions at higher levels and further space navigation, as symbolically shown in Figure 5.1.

Many spatial processes in SGL can start any time and in any place, cooperating or competing with each other, depending on applications. The self-spreading and self-matching SGL patterns-scenarios can create *active spatial infrastructures* covering any regions. These infrastructures can effectively support or express distributed knowledge bases, advanced command and control, situation awareness, autonomous and collective decisions, as well as any existing or hypothetical computational and/or control models, systems, and solutions. SGL has a deep recursive structure with its parallel spatial scenarios called *grasps* and universal operational and control constructs as *rules* (braces identifying repetition):

```
grasp → constant | variable | rule ({ grasp,})
```

5.3 INTEGRATING SATELLITE CONSTELLATIONS UNDER SGT

In the current state, as already mentioned in Chapter 2, practically all communication with satellites and between them is accomplished via ground-based antennas and infrastructures, as shown in Figure 5.2. These antennas are complicated for dealing with Low Earth Orbit (LEO) satellites as they

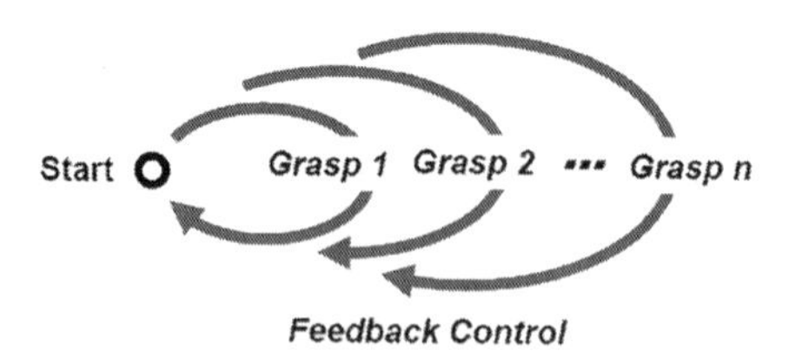

Figure 5.1 Controlled wavelike coverage and conquest of distributed spaces by SGT.

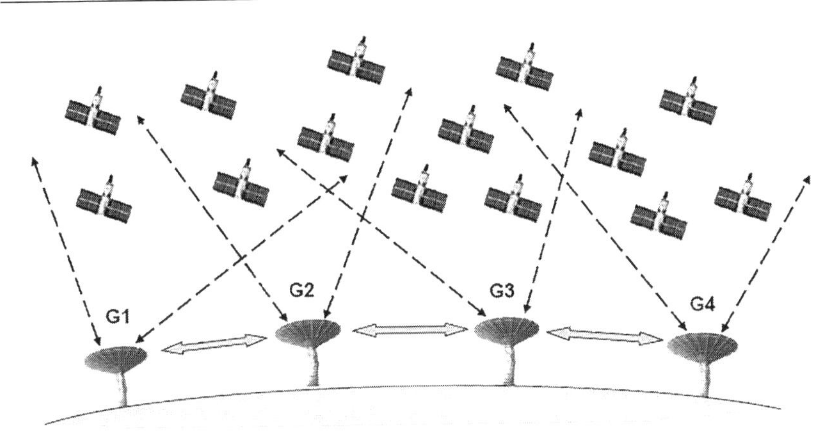

Figure 5.2 Communications with satellites and between them by ground antennas and infrastructures.

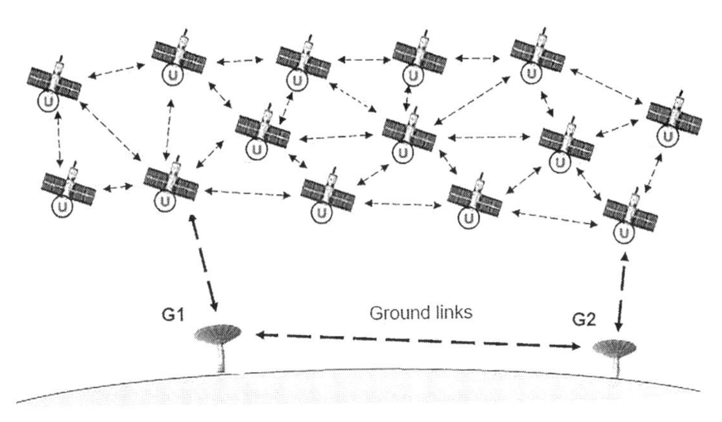

Figure 5.3 Converting the constellation into an integral system under SGT with simplification of ground infrastructures.

have to turn around and provide handover of loads between satellites which move fast around the globe.

But supplying satellites with SGL interpreters which may communicate directly with each other as an integral system and also with some ground stations may significantly simplify ground antennas and reduce their numbers, as shown in Figure 5.3.

Only a snapshot is shown in this figure, as the position of satellites and communication structure between them may rapidly change at runtime.

5.4 BROADCASTING EXECUTIVE ORDERS TO ALL SATELLITES

Starting from the first reached satellite from ground station `G1` and broadcasting via the dynamic network to all other satellites, with blocking possible cycling and delivering a given order for execution, as in Figure 5.4.

Each reached satellite first executes the order brought to it in frontal variable `Order` and then spreads further the network coverage via neighboring nodes, as by the following SGL scenario.

```
frontal(Order) = instructions;
hop(G1); hop_first_any(seen);
repeat(execute(Order); hop_first_all(seen))
```

The executing orders brought to satellites can also be processed by them in parallel with the spreading to other satellites, thus covering the whole constellation with executive orders and their execution much quicker, as follows.

```
frontal(Order) = instructions;
hop(G1); hop_first_any(seen);
repeat(free(execute(Order)), hop_first_all(seen))
```

Sending the order to a particular satellite `S10` via the dynamic constellation network, executing it, and then aborting the whole remaining network navigation can be expressed as follows.

```
frontal(Order = instructions, Destination = S10);
hop(G1); hop_first_any(seen);
repeat(if(NAME == Destination, (execute(Order); abort),
         hop_first_all(seen)))
```

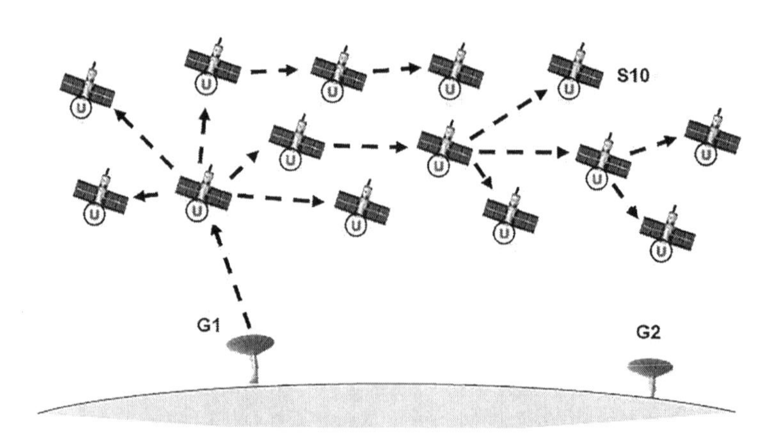

Figure 5.4 Broadcasting executive orders via parallel network navigation in a flooding mode.

5.5 BROADCASTING TO ALL SATELLITES WITH RETURNING THEIR ACCUMULATED DATA

Broadcasting to all satellites via their dynamic network from ground station G1, collecting the accumulated data (supposed to be in their personal nodal variables History), returning all this to G1 with further transference to another ground station G2, and fixing the final result there (see also Figure 5.5) may be achieved as follows.

```
hop(G1);
frontal(Summary) =
   (hop_first_any(seen);
    repeat(free(History), hop_first_all(seen)));
hop(G2); output(Summary)
```

If to collect histories from only particular satellite nodes (like S2, S5, S10), we may write in SGL:

```
hop(G1);
frontal(Summary) =
   (hop_first_any(seen);
    repeat(if(belong(NAME, (S2, S5, S10)), free(History));
    hop_first_all(seen)));
hop(ground, G2); output(Summary)
```

Real implementation of this scenario may depend on the constellation dynamics and stability of their network topology, where replying to the predecessor satellites with accumulated data may be complicated if their direct spatial links happen to be broken with passing time (say, when optical links used).

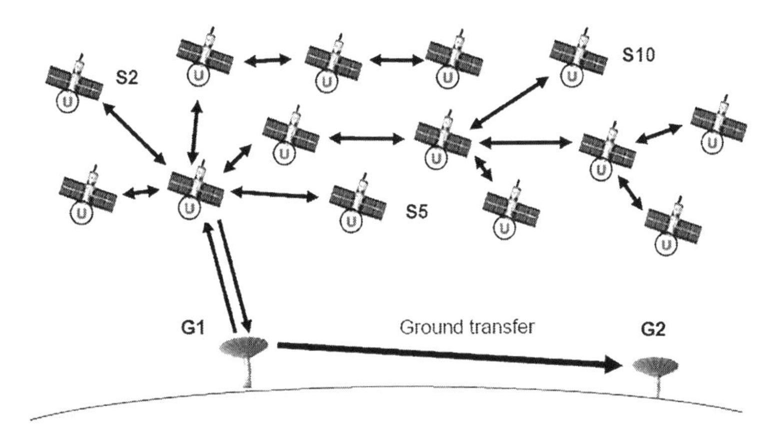

Figure 5.5 Broadcasting to the whole constellation with collection and return of the data accumulated in all satellites.

This may be done explicitly on the programming level for reaching a particular satellite via their current network topology (see Chapter 6 for this), where return to the previous satellite address automatically involves its search via navigation through the dynamic network.

5.6 CONSTELLATION REPOSITIONING AND RESTRUCTURING

Any needed constellation repositioning and restructuring, even emergency-like, can be done in a similar way, starting from any ground station, as shown in Figure 5.6 and by the following SGL scenario. This restructuring may involve removal from the orbits of used or malfunctioning satellites (hopefully not contributing to the rapidly growing huge collection of debris, see Chapter 9), and also adding new satellites to the constellation (not included in the scenario).

```
hop(G1);
frontal(Correction = advised, Removable = (S3, S6, S9));
hop(ground, G2); hop_first_any(seen);
repeat(
  free(update_position(History, Correction);
       if(belong(NAME, Removable), remove(current))),
  hop_first_all(seen))
```

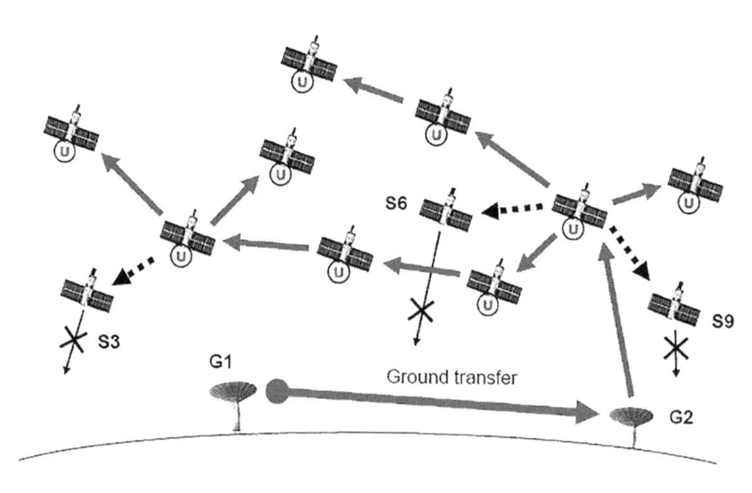

Figure 5.6 Constellation repositioning and restructuring with removal of used satellites.

5.7 TOWARD MORE COMPLEX CONSTELLATION SOLUTIONS UNDER SGT

In the previous sections we have considered only some basic operations on distributed satellite constellations, which can be organized as a whole by using the developed networking technology based on self-spreading and self-matching high-level recursive and parallel scenarios. In recent papers [27–34] we have shown some more specific operations on satellite networks under SGT, which may be useful for such important applications as global security and defense. These operations include effective use of organized satellite networks for tracking rapidly and complexly moving objects like hypersonic gliders, and constant observation of important objects on Earth in a custody-like manner, with low-flying LEO satellites, which are themselves rapidly moving around the globe and can be seen over its points for only a few minutes. We have also shown how using the SGT capability by operating in the same formalism with physical, virtual, and executive worlds as well as their combinations, we may essentially improve the capabilities of satellite constellations by introducing the virtual layer over satellite networks, which enables, for example, uninterrupted observation of any large infrastructures on Earth and not only local points or objects. In other papers [19–21], we also described using SGT for dealing with global terrestrial and celestial networks, where parallel spatial methods over large networks described there may also be useful for mega-constellations of satellites and their integration with ground-based networks and systems. More on these and other issues will be discussed in the following chapters of this book.

5.8 CONCLUSIONS

We have considered only some basic operations on distributed satellite constellations, which can be organized as a whole by using the developed networking technology based on self-spreading and self-matching high-level recursive and parallel scenarios. These include broadcasting of executive orders to all satellites, collecting and returning accumulated data by satellites, and constellation repositioning and restructuring which may be planned or in case of emergency. By supplying satellites with SGL interpreters which can directly communicate and cooperate with each other and also with the ground stations, we can convert the whole constellations into self-organized and, if needed, entirely space-managed systems with significant simplification of ground antennas and reduction of their numbers. More complex operations on satellite constellations including those related to different projects with them will be discussed in the subsequent chapters.

REFERENCES

1. R. Skibba, How satellite mega-constellations will change the way we use space, *MIT Technology Review*, February 26, 2020. https://www.technologyreview.com/2020/02/26/905733/satellite-mega-constellations-change-the-way-we-use-space-moon-mars/
2. M. Minet, The Space Legal Issues with Mega-Constellations, November 3, 2020. https://www.spacelegalissues.com/mega-constellations-a-gordian-knot/
3. E. Siegel, Astronomy Faces a Mega-Crisis as Satellite Mega-Constellations Loom, January 19, 2021. https://www.forbes.com/sites/startswithabang/2021/01/19/astronomy-faces-a-mega-crisis-as-satellite-mega-constellations-loom/?sh=30597dca300d
4. A. Jones, China is Developing Plans for a 13,000-Satellite Megaconstellation, *Space News*, April 21, 2021. https://spacenews.com/china-is-developing-plans-for-a-13000-satellite-communications-megaconstellation/
5. N. Reilanda, A. J. Rosengren, R. Malhotra, C. Bombardelli, Assessing and Minimizing Collisions in Satellite Mega-Constellations, 2021, Published by Elsevier B.V. on behalf of COSPAR. https://www.sciencedirect.com/science/article/abs/pii/S0273117721000326
6. J. Fous, Mega-Constellations and Mega-Debris, October 10, 2016. https://www.thespacereview.com/article/3078/1
7. N. Mohanta, How Many Satellites Are Orbiting the Earth in 2021? Geospatial World, 28 May, 2021. https://www.geospatialworld.net/blogs/how-many-satellites-are-orbiting-the-earth-in-2021/
8. United Nations Register of Objects Launched into Outer Space, The United Nations Office for Outer Space Affairs. http://www.unoosa.org/oosa/en/spaceobjectregister/index.html.
9. K. Dredge, M. von Arx, I. Timmins, LEO Constellations and Tracking Challenges, www.satellite-evolution.com, September/October 2017.
10. G. Curzi, D. Modenini, P. Tortora, Review Large Constellations of Small Satellites: A Survey of Near Future Challenges and Missions, *Aerospace* 7(9) (2020), 133. https://www.mdpi.com/2226-4310/7/9/133/htm
11. A. Venkatesan, J. Lowenthal, P. Prem, M. Vidaurri, The Impact of Satellite Constellations on Space as an Ancestral Global Commons, *Nature Astronomy* (4) 2020, November, 1043–1048. www.nature.com/natureastronomy
12. P.S. Sapaty, A distributed processing system, European Patent N 0389655, Publ. 10.11.93, European Patent Office. 35 p.
13. P.S. Sapaty, *Symbiosis of Real and Simulated Worlds under Spatial Grasp Technology*. Springer, 2021. 305 p.
14. P.S. Sapaty, *Complexity in International Security: A Holistic Spatial Approach*. Emerald Publishing, 2019. 160 p.
15. P.S. Sapaty, *Holistic Analysis and Management of Distributed Social Systems*. Springer, 2018. 234 p.
16. P.S. Sapaty, *Managing Distributed Dynamic Systems with Spatial Grasp Technology*. Springer, 2017. 284 p.
17. P.S. Sapaty, *Ruling Distributed Dynamic Worlds*. New York: John Wiley & Sons, 2005. 255 p.

18. P.S. Sapaty, *Mobile Processing in Distributed and Open Environments*. New York: John Wiley & Sons, 1999. 410 p.
19. P.S. Sapaty, Global Network Management under Spatial Grasp Paradigm. *International Robotics & Automation Journal*, 6(3), (2020) 134–148. https://medcraveonline.com/IRATJ/IRATJ-06-00212.pdf
20. P.S. Sapaty, Global Network Management under Spatial Grasp Paradigm, *Global Journal of Researches in Engineering: Journal of General Engineering* 20(5), 58–81 (2020). Version 1.0. https://globaljournals.org/GJRE_Volume20/6-Global-Network-Management.pdf
21. P.S. Sapaty, Advanced Terrestrial and Celestial Missions under Spatial Grasp Technology, *Aeronautics and Aerospace Open Access Journal* 4(3) (2020). https://medcraveonline.com/AAOAJ/AAOAJ-04-00110.pdf
22. P.S. Sapaty, Spatial Management of Distributed Social Systems, *Journal of Computer Science Research* 2(3), 2020, July. https://ojs.bilpublishing.com/index.php/jcsr/article/view/2077/pdf
23. P.S. Sapaty, Towards Global Nanosystems under High-level Networking Technology, *Acta Scientific Computer Sciences* 2(8) 2020. https://www.actascientific.com/ASCS/pdf/ASCS-02-0051.pdf
24. P.S. Sapaty, Symbiosis of Distributed Simulation and Control under Spatial Grasp Technology, *SSRG International Journal of Mobile Computing and Application (IJMCA)* 7(2) (2020, May–August). http://www.internationaljournalssrg.org/IJMCA/2020/Volume7-Issue2/IJMCA-V7I2P101.pdf
25. P.S. Sapaty, Symbiosis of Virtual and Physical Worlds under Spatial Grasp Technology, *Journal of Computer Science & Systems Biology* 13(6) (2020). https://www.hilarispublisher.com/open-access/symbiosis-of-virtual-and-physical-worlds-under-spatial-grasp-technology.pdf
26. P.S. Sapaty, Symbiosis of Real and Simulated Worlds under Global Awareness and Consciousness, The Science of Consciousness | TSC 2020. https://eagle.sbs.arizona.edu/sc/report_poster_detail.php?abs=3696
27. P.S. Sapaty Spatial Grasp as a Model for Space-based Control and Management Systems, *Mathematical Machines and Systems* 1 (2021), 135–138. http://www.immsp.kiev.ua/publications/articles/2021/2021_1/Sapaty_book_1_2021.pdf
28. P.S. Sapaty, Managing Multiple Satellite Architectures by Spatial Grasp Technology, *Mathematical Machines and Systems* 1 (2021), 3–16. http://www.immsp.kiev.ua/publications/eng/2021_1/
29. P.S. Sapaty, Spatial Management of Large Constellations of Small Satellites, *Mathematical Machines and Systems* 2 (2021). http://www.immsp.kiev.ua/publications/articles/2021/2021_2/02_21_Sapaty.pdf
30. P.S. Sapaty, Global Management of Space Debris Removal under Spatial Grasp Technology, *Acta Scientific Computer Sciences* 3(7) (2021, July). https://www.actascientific.com/ASCS/pdf/ASCS-03-0135.pdf
31. P.S. Sapaty, Space Debris Removal under Spatial Grasp Technology, *Network and Communication Technologies* 6(1) (2021). https://www.ccsenet.org/journal/index.php/nct/article/view/0/45486
32. P.S. Sapaty, Spatial Grasp Model for Management of Dynamic Distributed Systems, *Acta Scientific Computer Sciences* 3(9) (2021). https://www.actascientific.com/ASCS/pdf/ASCS-03-0170.pdf

33. P.S. Sapaty, Spatial Grasp Model for Dynamic Distributed Systems, *Mathematical Machines and Systems* 3 (2021). http://www.immsp.kiev.ua/publications/articles/2021/2021_3/03_21_Sapaty.pdf
34. P.S. Sapaty, Development of Space-based Distributed Systems under Spatial Grasp Technology, *Mathematical Machines and Systems* 4 (2021), 3–14.

Chapter 6

Transport layer organization under Spatial Grasp Technology (SGT)

6.1 INTRODUCTION

The current chapter discusses how the developed high-level system philosophy and model can effectively organize distributed space-based systems on different stages of their development and growth. The Spatial Grasp Technology (SGT), described in Chapters 3 and 4, based on parallel pattern-matching of distributed environments with high-level recursive mobile code can effectively provide any networking protocols and important applications of large satellite constellations, especially those on Low Earth Orbits (LEOs).

The chapter contains examples of technology-based solutions for establishing basic communications between satellites, starting from their initial, often chaotic, launches and distributing and collecting data in the growing constellations with even unstable and rapidly changing connections between satellites. It describes how to organize and register networking topologies in case of predictable distances between satellites, and how the fixed networking structures can help in solving complex problems. The chapter is based on analysis of existing publications related to satellite constellations and planned projects with them [1–27], especially related to transport networks and routing [19–27], also on the experience obtained from the development of SGT and the previous technology versions, as well as from their many researched and tested applications [28–51].

The rest of this chapter is organized as follows. *Section 6.2* reminds a brief summary of SGT described in detail in Chapters 3 and 4. *Section 6.3* briefs on Space Development Agency (SDA) transport layer aimed at providing basic satellite communications and their networking for the new space architecture. *Section 6.4* shows how to deal with highly dynamic constellation topologies under SGT, which includes reaching proper satellite nodes, delivery of a package to proper nodes, reaching proper destinations with the return and collection of certain items from them, and introducing special measures for dealing with high unpredictability of topology dynamics. *Section 6.5* deals with working with stable constellation topologies while

DOI: 10.1201/9781003230090-6

explaining advantages of them, creating and fixing communication topology in case of stable distances between satellites, and expressing in Spatial Grasp Language (SGL) such basic networking operations as finding shortest paths and routing tables. *Section 6.6* shows how to find certain structures and components in distributed networks with stable topologies, like articulation points and cliques. In *Section 6.7*, a routine satellite constellation management example is shown under stable topologies, which provides collection and return of remote data from the proper destinations. *Section 6.8* concludes the chapter.

6.2 A BRIEF SUMMARY OF SPATIAL GRASP TECHNOLOGY

Within SGT, described in Chapters 3–4, also in more detail in [28–34], a high-level scenario for any task to be performed in a distributed world is represented as an *active self-evolving pattern* rather than a traditional program, sequential or parallel one. This pattern, written in a recursive high-level SGL and expressing top semantics of the problem to be solved, can start from any point of the world. Then it spatially *propagates, replicates, modifies, covers, and matches* the distributed world in a parallel wavelike mode, while echoing the reached control states and data found or obtained for making decisions at higher levels and further space navigation, as symbolically shown in Figure 6.1.

Many spatial processes in SGL can start any time and in any place, cooperating or competing with each other, depending on applications. The self-spreading and self-matching SGL patterns-scenarios can create *active spatial infrastructures* covering any region. These infrastructures can effectively support or express distributed knowledge bases, advanced command and control, situation awareness, autonomous and collective decisions, as well as any existing or hypothetical computational and/or control models, systems, and solutions. SGL has a deep recursive structure with its parallel spatial scenarios called *grasps* and universal operational and control constructs called *rules* (braces identifying repetition):

```
grasp → constant | variable | rule ({ grasp,})
```

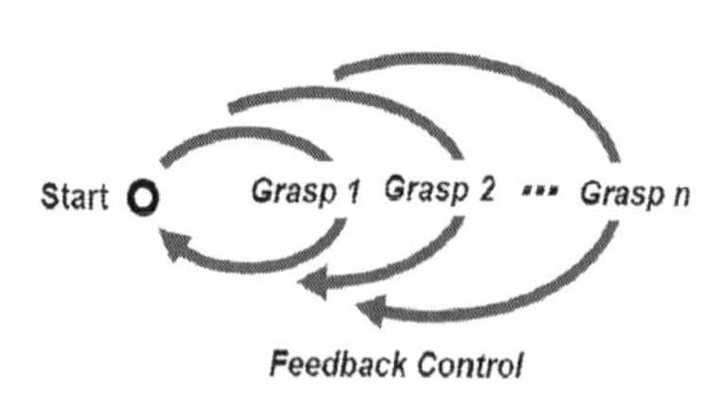

Figure 6.1 Controlled wavelike coverage and conquest of distributed spaces by SGT.

6.3 SPACE DEVELOPMENT AGENCY (SDA) TRANSPORT LAYER

The SDA's immediate goal is the development of Transport Layer [19–21], see Figure 6.2, consisting of a mesh network for communications and data in LEO that will talk to each other and relay data to military forces on the ground. The transport layer will provide assured, resilient, low-latency military data and connectivity worldwide to the full range of warfighter platforms.

It is envisioned, modeled, and architected as a constellation varying in size from 300 to more than 500 satellites in LEO ranging from 750 km to 1,200 km in altitude. With a full constellation, 95% of the locations on the Earth will have at least two satellites in view at any given time while 99% of the locations on the Earth will have at least one satellite in view.

This will ensure constant worldwide coverage around the globe. The constellation will be interconnected with Optical Intersatellite Links (OISLs), which have significantly increased performance over existing radio frequency crosslinks. LEO orbits in conjunction with OISLs will reduce path loss issues but more importantly offer much lower latencies, which are deemed critical to prosecute time-sensitive targets in today's wartime environment.

6.4 DEALING WITH DYNAMIC CONSTELLATION TOPOLOGIES

In general, constellations may consist of satellites on different orbits, which may be changing in time their distances to each other, thus having no stable network topologies, as symbolically shown in Figure 6.3. But we will assume here that there are enough of them around the Earth, and at each moment of time any satellite can directly communicate with other satellites (at least one of them), which may be very different from the previous communications. So this in principle could enable to propagate through the whole constellation regardless of which satellites are now visible from each other.

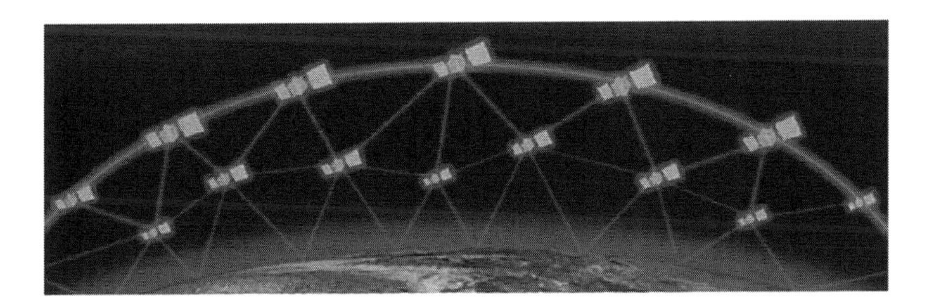

Figure 6.2 SDA transport layer.

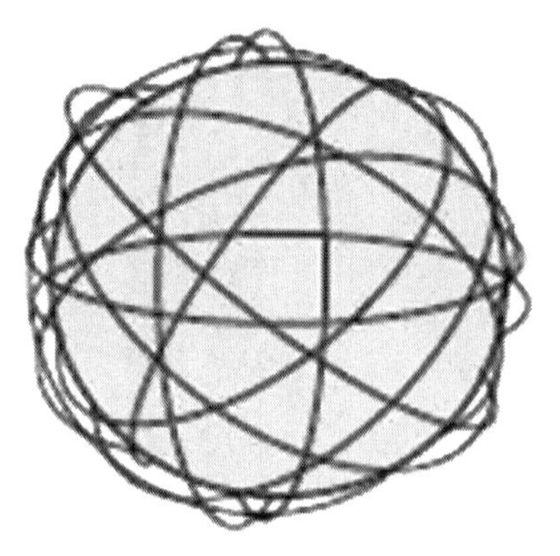

Figure 6.3 An example of a multiple and complex orbit satellite constellation.

We will consider some basic operations on such dynamic networks under SGT, expressed in SGL.

6.4.1 Reaching proper satellite nodes

a) Reaching particular satellite node `a` (like one of nodes of Figure 6.4) from a ground station initially via an arbitrary initial satellite visible from it at this moment of time can be accomplished as follows, with subsequent propagation via other satellites (if this starting node is not `a` itself). The satellites of interest are supposed to be of certain `Type`, and direct communication between them can be possible by some `Depth` threshold distance. The reached satellite is supposed to acknowledge the fact being reached by a symbolic `output` operation,

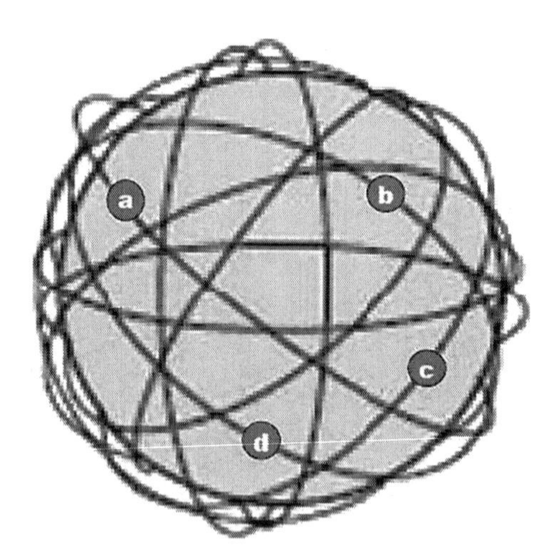

Figure 6.4 Certain nodes to be reached in complex satellite constellation.

after which propagation to other satellites will be canceled by the done statement. The operation hopfirst by reaching any new satellite node properly marks it, blocking a repeated entry and therefore any navigation cycling, which could become endless otherwise.

```
frontal(Depth = …, Type = …);
hopfirst(any, Type);
repeat(
  if(NAME == a, done(output(ok)));
  hopfirst(Depth, Type))
```

b) Reaching particular satellite node a from another satellite node after initially staying in it, without involvement of ground stations, can be a bit simpler, as follows.

```
frontal(Depth = …, Type = …);
repeat(
  if(NAME == a, done(output(ok)));
  hopfirst(Depth, Type))
```

c) Reaching particular group of nodes from another satellite node after initially staying in it, where the found nodes may also serve as transits to the other Dest nodes sought.

```
frontal(Depth = …, Type = …, Dest = (a, b, c, d));
repeat(
  if(belong(NAME, Dest), output(ok));
  hopfirst(Depth, Type))
```

This can also be accomplished from a ground station getting accidental access to any satellite node, similar to the first case above.

```
frontal(Depth = …, Type = …, Dest = (a, b, c, d));
hopfirst(any, Type);
repeat(
  if(belong(NAME, Dest), output(ok));
  hopfirst(Depth, Type))
```

We may also use spatial recursion instead of spatial cycling for the previous case, while navigating and covering the system in parallel too, as follows.

```
frontal(
  Depth = …, Type = …, Dest = (a, b, c, d));
  Cover = {if(belong(NAME, Dest), output(ok));
           hopfirst(Depth, type); run(Cover)};
hopfirst(any, Type); run(Cover)
```

6.4.2 Package delivery to given destinations

Reaching the set of particular nodes with delivering a package into them all from the ground station, using spatial cycling and taking for simplicity that this package will be attached to their contents (the latter expressed by standard nodal environmental variable CONTENT).

```
frontal(Depth = …, Type = …, Pack = …,
        Dest = (a, b, c, d));
hopfirst(any, Type);
repeat(
  if(belong(NAME, Dest), append(Pack, CONTENT));
  hopfirst(Depth, Type))
```

The same on the result, but with the use of spatial recursion:

```
frontal(
  Depth = …, Type = …, Dest = (a, b, c, d), Pack = …;
  Deliver = {if(belong(NAME, Dest), append(Pack, CONTENT));
             hopfirst(Depth, type); run(Deliver)};
hopfirst(any, Type); run(Deliver)
```

For the both cases above, instead of CONTENT any other variables could be used in the destination nodes. Also, the delivery may be initially launched from any satellite node.

6.4.3 Moving to proper destinations and returning packs from them

We will consider first picking up a concrete Item in all destinations after reaching them (supposedly from their CONTENT), and then their return, collection, and output from the casual starting satellite node to the ground station. From the very beginning, mobile environmental variable IDENTITY has default value like the starting node name, address (if has it), or just randomly generated value (with which all subsequently reached nodes will be marked by hopfirst, unless this value is changed explicitly, as in the following example for the backward propagations).

```
frontal(Depth = …, Type = …, Item = …, Pack, Start,
        Dest = (a, b, c, d));
hopfirst(any, Type); nodal(Summary), Start = NAME);
sequence(
 repeat(
   if(belong(NAME, Dest),
      free(IDENTITY = NAME;
           Pack = element(CONTENT, Item);
           repeat(if(NAME == Start,
                     blind(append(Summary, Pack)));
                  hopfirst(Depth, Type))));
   hopfirst(Depth, Type))),
 output(Summary))
```

By output, this pack collection may be delivered to the ground station. The collected packs will also remain in variable Summary in the starting satellite node. The above scenario combines forward and backward simultaneous spatial processes, where the forward one spreads from the starting node to reach the needed destinations, and the backward ones immediately start in all reached nodes to deliver their items to the starting node, without waiting the completion of forward process.

Another solution may be with the use of spatial coverage recursion, as follows (for reaching the needed nodes but not for returning results, which remain spatially repetitive).

```
frontal(
  Depth = …, Type = …, Item = …, Pack, Start, Dest = (a, b,
c, d),
  Cover = {(belong(NAME, Dest); IDENTITY = NAME;
             Pack = element(CONTENT, Item);
             repeat(if(NAME == Start, blind(append(Summary,
Pack)));
                       hopfirst(Depth, Type))),
           (hopfirst(Depth, Type); apply(Cover))};
hopfirst(any, Type); nodal(Summary); Start = NAME;
sequence(apply(Cover), output(Summary))
```

6.4.4 Providing further flexibility for dealing with dynamic topologies

We will consider here the possibility of reaching particular nodes from a ground station by allowing nodes to be entered more than once, assuming that with passing time other nodes may become directly accessible from the same nodes. The parameter Revisit shows how many times the nodes can be revisited. The sought destinations may be ignored when tried to be reentered, as representing the results already reported, for which the mark Visited is used in them.

```
frontal(Depth = …, Type = …, Dest = (a, b, c, d), Revisit = 3);
nodal(Visited);
hopfirst(Revisit)(any, Type);
repeat(
  if(belong(NAME, Dest),
     done(if(Visited == nil, (Visited = 1; output(ok)))));
  hopfirst(Revisit)(Depth, Type))
```

This hopfirst option is also using the quality of hopforth rule not allowing to reenter the nodes just visited before them. The discussed nodes revisiting variant can also be used in all previous scenario examples for cases with very high dynamics and unpredictability of spatial distribution of satellite constellations.

6.5 WORKING WITH STABLE CONSTELLATION TOPOLOGIES

With established and registered stable links between nodes (assumed here for simplicity, all having length 1) and with stable network topology (like the one in Figure 6.5) we can solve different problems in satellites constellations much easily.

6.5.1 Creating stable constellation network topology and its advantages

Starting from a ground station and linking certain `Type` nodes close enough in space, we may obtain a stable and useful network topology (using `Depth` threshold distance between satellites reflecting their capability of direct communication and official linking with each other). Such network topology can be created in parallel with the constellation navigation, which can also be done in parallel itself (all established links between neighboring satellites are just named as `link`), as follows.

```
frontal(Depth = …, Type = …);
hopfirst(any, Type);
repeat(
  free(hop_nodes(Depth, Type); if(no_links(BACK),
linkup(link, BACK)));
  hopfirst(Depth, Type))
```

By getting a stable topology, we can use the full power of SGT and its basic language SGL for solving any problems on distributed satellite networks, with numerous publications already published in this area, including [29–51]. A few of the problems are finding shortest paths, routing tables, weak and strong components like articulation points and cliques, any particular structures and components (say, represented as spatial images or patterns),

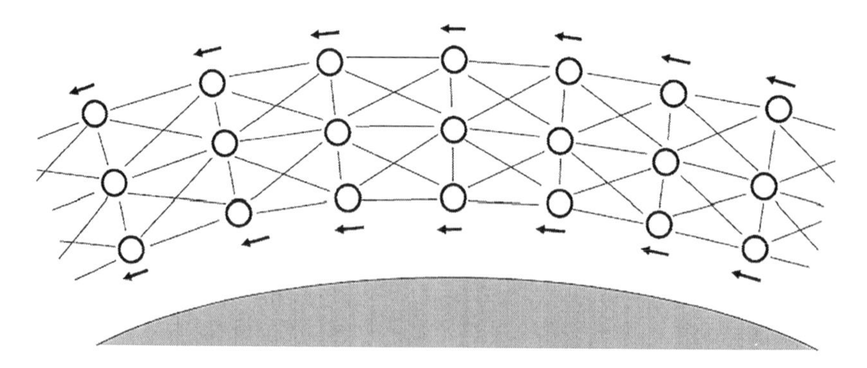

Figure 6.5 Example of a stable constellation network.

and so on. SGT and SGL may be very useful for large constellations management—from creation of celestial version of internet—to weather prediction—to global industry—to security and defense. Some examples of advantages of stable topologies follow.

6.5.2 Finding shortest path tree from a node to all other nodes

Such a tree from some node a with all network links having symbolic length 1 may be as the one shown in Figure 6.6(a) (another possible SPT for this network and from the same node in shown in Figure 6.6(b)).

The following scenario, starting with some node, let it be a, can accomplish this with registering the resultant SPT in variables Up in the network nodes. Such variables are reflecting the tree higher (or predecessor) nodes, and are regularly updated if a better solution for the shortest path to these nodes emerges during parallel network navigation.

```
nodal(Distance, Up); frontal(Far);
hop(a); Distance = 0;
repeat(
  hop_links(all); increment(Far, 1);
  or(Distance == nil, Distance > Far);
  Distance = Far; Up = BACK)
```

6.5.3 Collecting shortest path via Shortest Path Tree (SPT)

Collecting and reporting the shortest path from any other node like f to the starting node, beginning from f and issued in a, will be as follows:

```
hop(f); frontal(Path);
repeat(Path = append(Path, NAME); hop(Up));
output(Path)
```

The result will be as: (f, d, a)

If the path to be presented in the opposite order, i.e., from a to f and issued in a too, we will have;

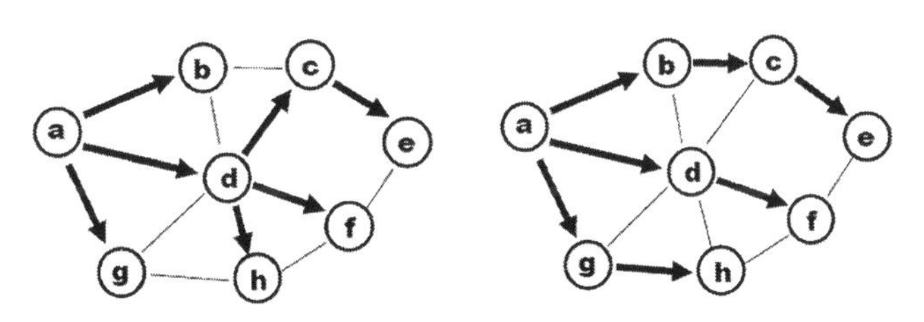

Figure 6.6 Possible shortest path trees from a node to all other nodes.

```
hop(f); frontal(Path);
repeat(Path = append(NAME, Path); hop(Up));
output(Path)
```

The result will be: `(a, d, f)`

If we decide to collect the shortest path starting from the top SPT node a to other node `f`, and finally report the result in a too, we will have:

```
hop(a); frontal(Path) = NAME;
output(
  repeat(hop_links(all); Up == BACK;
         Path = append(Path, NAME);
         if(NAME == f, done(Path))))
```

If we decide to collect the path starting in a and report it in final node `f`, we should write:

```
hop(a); frontal(Path) = NAME;
repeat(hop_links(all); Up == BACK;
       Path = append(Path, NAME);
       if(NAME == f, done(output(Path))))
```

6.5.4 Creating routing tables

Parallel SGL scenarios for finding routing tables for distributed networks can be found in [29–31]. An example of such tables for all nodes of the network depicted in Figure 6.6 is shown in Figure 6.7, where the `Dest` vector in each node collects names of all other nodes as final destinations, and the `Next` vector—the next nodes through which these destinations can be reached via shortest paths to them.

For example, finding the shortest path from any node, say `g`, to any other node, like e, via the routing tables of Figure 6.7 can be organized in SGL as follows:

```
hop(g); repeat(hop(link(any), node(element(Next, order(Dest,
e)))))
```

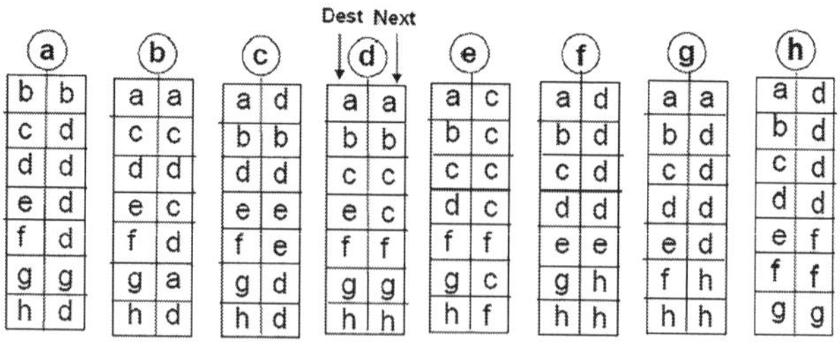

Dest Next

a		b		c		d		e		f		g		h	
b	b	a	a	a	d	a	a	a	c	a	d	a	a	a	d
c	d	c	c	b	b	b	b	b	c	b	d	b	d	b	d
d	d	d	d	d	d	c	c	c	c	c	d	c	d	c	d
e	d	e	c	e	e	e	c	d	c	d	d	d	d	d	d
f	d	f	d	f	e	f	f	f	f	e	e	e	d	e	f
g	g	g	a	g	d	g	g	g	c	g	h	f	h	f	f
h	d	h	d	h	d	h	h	h	f	h	h	h	h	g	g

Figure 6.7 Full collection of routing tables in all nodes for the network of Figure 6.6.

This scenario will follow the path g→d→c→e with using the tables of Figure 6.7 in nodes g, d, and c.

6.6 FINDING PARTICULAR COMPONENTS AND STRUCTURES IN STABLE NETWORKS

In dealing with large distributed systems, including satellite constellations, of particular interest may be finding certain features or structures in them (like, for example, their weak and strong points or components), which may be useful for solving different optimization and security problems. We will provide below some related examples in SGL taking into account that the satellite constellation is reasonably stable and structured. More solutions with detailed explanations can be found in [29–34, 39, 40].

6.6.1 Finding weakest points

An example of weakest points in networked systems may be their articulation points (two of them are shown in Figure 6.8) which, when removed, each separately, split the network into disjoint parts.

The following scenario finds all such nodes in a parallel and distributed mode, resulting in the output of node names d and c for the network of Figure 6.8.

```
hop_nodes(all); IDENTITY = NAME; hopfirst(current);
and_sequence(
  (hopfirst(random(link, any)); repeat(hopfirst(links,
all))),
  hopfirst_links(all),
  output(NAME))
```

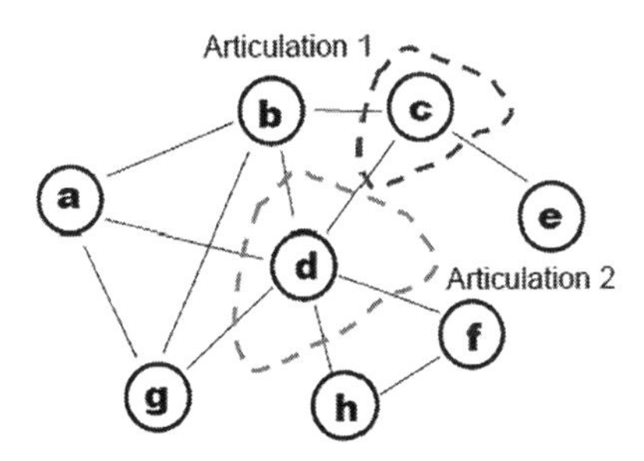

Figure 6.8 Network articulation points.

6.6.2 Finding strongest points

In this respect, cliques (or maximum fully connected sub-graphs of a graph, as in Figure 6.9) are often considered as the strongest parts of a network.

They all can be found in parallel by the following simple scenario (establishing lower threshold on the number of fully interconnected nodes as 3 to be considered as a clique):

```
Hop_nodes(all); frontal(Clique = NAME);
repeat(
  hop_links(all); notbelong(NAME, Clique);
  if(and_parallel(hop(links(any), all_nodes(Clique))),
     if(BACK > NAME, append(NAME, Clique), done),
     fail));
count(Clique) >= 3; output(Clique)
```

It will result in the following cliques for the network of Figure 6.9:

```
(a, b, d, g), (b, c, d), (d, f, h)
```

6.6.3 Some comments on finding different structures in networks

Having found the weakest points in a distributed network, we can impact it in different ways, for example, by removing these points to weaken the network further, or by adding new links between other nodes, especially its direct neighbors, to strengthen it. After finding cliques, we can organize a certain impact on the network, say, by weakening it via removing of some links or/and nodes in it. This may result in the network with reduced size of cliques, lesser number of them, or without cliques at all. In general, we can easily find in SGL any particular structures and components (say, represented as spatial images or patterns) and in arbitrary networks, see many details in [29–34]. But for doing this most efficiently and in suitable time, the constellation network structures and established links between nodes should remain stable, at least as much as possible.

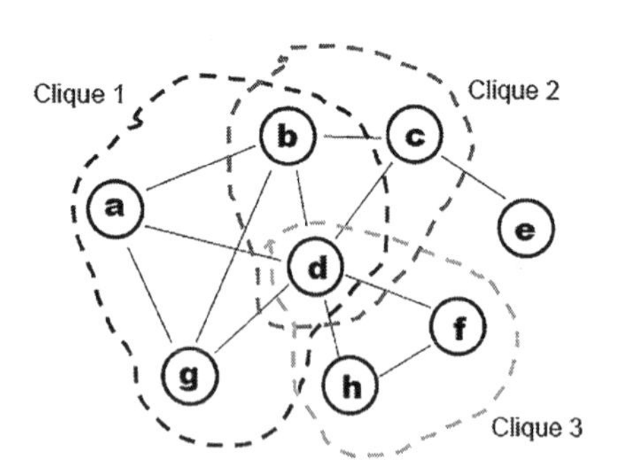

Figure 6.9 Strongest parts, or cliques.

6.7 EXAMPLE OF USING STABLE CONSTELLATION NETWORKS

As an elementary example of advantage of such stable network topology, we may show how simple it will be picking up certain items in the given destinations and their return, collection, and output at the ground station, using registered paths from the start to destinations in a reverse order. (See for comparison the related previous examples for such a task in Section 6.4 but with unpredictable and rapidly changeable constellation structures.)

```
frontal(Item = …, Dest = (a, b, c, d));
output(
   hopfirst_node(any);
   repeat(if(belong(NAME, Dest),
             free(element(CONTENT, Item)));
          hopfirst_links(all)))
```

This may be even simpler if we introduce such possibility as direct hops to given nodes by their names, which can be easily achieved with the use of routing tables (see above) providing shorted paths to the needed nodes from the starting satellite. And then these paths after being registered in the network structure are providing echoing and collecting proper items from the reached needed destinations and outputting them altogether in the starting node (see also Figure 6.10).

```
frontal(Item) = …; Nodal(Dest) = (a, b, c, d);
output(hop_nodes(Dest); element(CONTENT, Item))
```

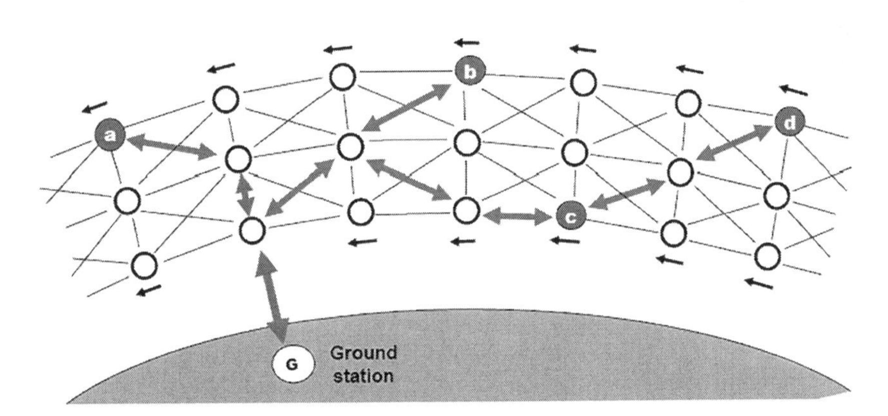

Figure 6.10 Collection and return of remote data in a stable constellation network.

6.8 CONCLUSIONS

This chapter was devoted to the transport layer, including the one of SDA architecture [20], which is the basic and most important layer for any multiple satellite systems, as providing main communications between satellites and enabling them to be used as a system in different projects. The chapter offered a unified approach toward organization of large satellite constellations, especially on Low Earth Orbits, and on any stages of their development, starting from their accidental launches and gradual growth. Such constellations at the beginning may not have stable communication structures and network topologies similar to the existing terrestrial systems. All exemplary solutions and on different levels were presented in the same Spatial Grasp Language which with its recursive grasping and matching of distributed spaces appeared to be suitable and universal for the whole constellation management. The chapter shows how to deal with highly dynamic and even unpredictable constellation topologies, which included reaching proper satellite nodes, delivery of packages to proper nodes, and reaching proper destinations with the return and collection of certain items from them. It then showed how to create stable constellation networks in case of fixed distances between satellites, and demonstrated advantages of using them, including finding shortest paths between satellites, routine tables, and discovering different components and structures in networks which may be useful in advanced constellation management projects.

REFERENCES

1. N. Mohanta, How Many Satellites Are Orbiting the Earth in 2021? *Geospatial World*, 05/28 (2021, May). https://www.geospatialworld.net/blogs/how-many-satellites-are-orbiting-the-earth-in-2021/
2. United Nations Register of Objects Launched into Outer Space, *The United Nations Office for Outer Space Affairs*. http://www.unoosa.org/oosa/en/spaceobjectregister/index.html
3. Martin G. NewSpace: The «Emerging» Commercial Space Industry. https://ntrs.nasa.gov/archive/nasa/casi.ntrs.nasa.gov/20140011156.pdf
4. Bockel J.-M. The Future of the Space Industry. *General Report*, November 17, 2018.https://www.nato-pa.int/download-file?filename=sites/default/files/2018-12/2018%20-%20THE%20FUTURE%20OF%20SPACE%20INDUSTRY%20-%20BOCKEL%20REPORT%20-%20173%20ESC%2018%20E%20fin.pdf
5. G. Curzi, D. Modenini, P. Tortora, Review large constellations of small satellites: A survey of near future challenges and missions, *Aerospace* 7(9) (2020), 133. https://www.mdpi.com/2226-4310/7/9/133/htm
6. A. Venkatesan, J. Lowenthal, P. Prem, M. Vidaurri, The impact of satellite constellations on space as an ancestral global commons, *Nature Astronomy* 4 (2020, November), 1043–1048. www.nature.com/natureastronomy

7. R. Skibba, How satellite mega-constellations will change the way we use space, *MIT Technology Review*, February 26 (2020). https://www.technologyreview.com/2020/02/26/905733/satellite-mega-constellations-change-the-way-we-use-space-moon-mars/
8. M. Minet, The Space Legal Issues with Mega-Constellations, November 3, 2020. https://www.spacelegalissues.com/mega-constellations-a-gordian-knot/
9. E. Siegel, Astronomy Faces a Mega-Crisis as Satellite Mega-Constellations Loom, January 19, 2021. https://www.forbes.com/sites/startswithabang/2021/01/19/astronomy-faces-a-mega-crisis-as-satellite-mega-constellations-loom/?sh=30597dca300d
10. A. Jones, China is developing plans for a 13,000-satellite megaconstellation, *Space News*, April 21, 2021. https://spacenews.com/china-is-developing-plans-for-a-13000-satellite-communications-megaconstellation/
11. N. Reilanda, A. J. Rosengren, R. Malhotra, C. Bombardelli, Assessing and minimizing collisions in satellite mega-constellations, 2021, Published by Elsevier B.V. on behalf of COSPAR. https://www.sciencedirect.com/science/article/abs/pii/S0273117721000326
12. J. Fous, Mega-Constellations and Mega-Debris, October 10, 2016. https://www.thespacereview.com/article/3078/1
13. Space Development Agency Next-Generation Space Architecture. 2019. https://www.airforcemag.com/PDF/DocumentFile/Documents/2019/SDA_Next_Generation_Space_Architecture_RFI%20(1).pdf.
14. S. Magnuson, Web Exlusive: Details of the Pentagon's New Space Architecture Revealed. 2019. https://www.nationaldefensemagazine.org/articles/2019/9/19/details-of-the-pentagon-new-space-architecture-revealed.
15. D. Messier, Space Development Agency Seeks Next-Gen Architecture in First RFI. 2019. http://www.parabolicarc.com/2019/07/07/space-development-agency-issues-rfi/
16. V. Insinna, Space Agency has an Ambitious Plan to Launch 'Hundreds' of Small Satellites. Can it get off the ground? *DefenceNews*. Space. 2019. April 10. https://www.defensenews.com/space/2019/04/10/sda-has-an-ambitious-plan-to-launch-hundreds-of-small-satellites-can-it-get-off-the-ground/
17. N. Strout, Space Development Agency Approves Design for Satellites That Can Track Hypersonic Weapons, *C4ISRNET*, 2021. https://www.c4isrnet.com/battlefield-tech/space/2021/09/20/space-development-agency-approves-design-for-satellites-that-can-track-hypersonic-weapons/
18. S. Erwin, Space Development Agency could select three manufacturers to produce its next batch of satellites, *Space News*, 2021. https://spacenews.com/space-development-agency-could-select-three-manufacturers-to-produce-its-next-batch-of-satellites/
19. Transport Layer Tranche-0 A-Class 1, ..., 7. https://space.skyrocket.de/doc_sdat/transport-layer-tranche-0-a-class-york.htm
20. Transport, SDA. https://www.sda.mil/transport/
21. Transport Layer. https://en.wikipedia.org/wiki/Transport_layer
22. M.A.A. Madni, S. Iranmanesh, R. Raad, Review DTN and Non-DTN Routing Protocols for Inter-CubeSat Communications: A comprehensive survey. *Electronics* 9 (2020), 482. https://www.mdpi.com/2079-9292/9/3/482

23. Q.I. Xiaogang, M.A. Jiulong, W.U. Dan, L.I.U. Lifang, Shaolin H.U. A survey of routing techniques for satellite networks. *Journal of Communications and Information Networks* 1(4) (2016), 66–85. http://www.infocomm-journal.com/jcin/EN/10.11959/j.issn.2096-1081.2016.058.
24. M.A.A. Madni, S. Iranmanesh, R. Raad, Review DTN and Non-DTN Routing Protocols for Inter-CubeSat Communications: A comprehensive survey. *Electronics* 9 (2020), 482. https://www.mdpi.com/2079-9292/9/3/482
25. L. Wood, Satellite Constellation Networks, *ResearchGate*, 2003. https://www.researchgate.net/publication/2559727_Satellite_Constellation_Networks
26. Y. Ma, et al., A Distributed Routing Algorithm for LEO Satellite Networks, 2013, IEE Digital library, https://www.computer.org/csdl/proceedings-article/trustcom/2013/5022b367/12OmNqOffvl
27. J. H. Bau, *Topologies for Satellite Constellations in a Cross-linked Space Backbone Network*, Massachusetts Institute of Technology, 2000. https://dspace.mit.edu/bitstream/handle/1721.1/38441/51111999-MIT.pdf?sequence=2&isAllowed=y
28. P.S. Sapaty, A distributed processing system, European Patent N 0389655, Publ. 10.11.93, European Patent Office. 35 p.
29. P.S. Sapaty, *Symbiosis of Real and Simulated Worlds under Spatial Grasp Technology*. Springer, 2021. 305 p.
30. P.S. Sapaty, *Complexity in International Security: A Holistic Spatial Approach*. Emerald Publishing, 2019. 160 p.
31. P.S. Sapaty, *Holistic Analysis and Management of Distributed Social Systems*. Springer, 2018. 234 p.
32. P.S. Sapaty, *Managing Distributed Dynamic Systems with Spatial Grasp Technology*. Springer, 2017. 284 p.
33. P.S. Sapaty, *Ruling Distributed Dynamic Worlds*. New York: John Wiley & Sons, 2005. 255 p.
34. P.S. Sapaty, *Mobile Processing in Distributed and Open Environments*. New York: John Wiley & Sons, 1999. 410 p.
35. P.S. Sapaty, Advanced terrestrial and celestial missions under spatial grasp technology. *Aeronautics and Aerospace Open Access Journal* 4(3) (2020). https://medcraveonline.com/AAOAJ/AAOAJ-04-00110.pdf
36. P.S. Sapaty, Spatial management of distributed social systems, *Journal of Computer Science Research* 2(3) (2020). https://ojs.bilpublishing.com/index.php/jcsr/article/view/2077/pdf
37. P.S. Sapaty, Towards Global Nanosystems under High-level Networking Technology, *Acta Scientific Computer Sciences* 2(8) (2020). https://www.actascientific.com/ASCS/pdf/ASCS-02-0051.pdf
38. P.S. Sapaty, Symbiosis of Distributed Simulation and Control under Spatial Grasp Technology, *SSRG International Journal of Mobile Computing and Application (IJMCA)* 7(2) (2020, May–August). http://www.internationaljournalssrg.org/IJMCA/2020/Volume7-Issue2/IJMCA-V7I2P101.pdf
39. P.S. Sapaty, Global Network Management under Spatial Grasp Paradigm, *International Robotics & Automation Journal* 6(3) (2020). https://medcraveonline.com/IRATJ/IRATJ-06-00212.pdf
40. P.S. Sapaty, Global Network Management under Spatial Grasp Paradigm, *Global Journal of Researches in Engineering: Journal of General Engineering*

20(5) (2020). https://globaljournals.org/GJRE_Volume20/6-Global-Network-Management.pdf
41. P.S. Sapaty, Symbiosis of Real and Simulated Worlds under Global Awareness and Consciousness, *The Science of Consciousness Symposium TSC* (2020). https://eagle.sbs.arizona.edu/sc/report_poster_detail.php?abs=3696
42. P.S. Sapaty, Fighting global viruses under spatial grasp technology, *Transactions on Engineering and Computer Science* 1(2) (2020). https://gnoscience.com/uploads/journals/articles/118001716716.pdf
43. P.S. Sapaty, Symbiosis of Virtual and Physical Worlds under Spatial Grasp Technology, *Journal of Computer Science & Systems Biology* 13(6) (2020). https://www.hilarispublisher.com/open-access/symbiosis-of-virtual-and-physical-worlds-under-spatial-grasp-technology.pdf
44. P.S. Sapaty, Spatial Grasp as a Model for Space-based Control and Management Systems, *Mathematical Machines and Systems* 1 (2021), 135–138. http://www.immsp.kiev.ua/publications/articles/2021/2021_1/Sapaty_book_1_2021.pdf
45. P.S. Sapaty, Managing multiple satellite architectures by spatial grasp technology, *Mathematical Machines and Systems* 1 (2021), 3–16. http://www.immsp.kiev.ua/publications/eng/2021_1/
46. P.S. Sapaty, Spatial Management of Large Constellations of Small Satellites, *Mathematical Machines and Systems* 2 (2021). http://www.immsp.kiev.ua/publications/articles/2021/2021_2/02_21_Sapaty.pdf
47. P.S. Sapaty, Global Management of Space Debris Removal under Spatial Grasp Technology, *Acta Scientific Computer Sciences* 3(7) (2021, July). https://www.actascientific.com/ASCS/pdf/ASCS-03-0135.pdf
48. P.S. Sapaty, Space Debris Removal under Spatial Grasp Technology, *Network and Communication Technologies* 6(1) (2021). https://www.ccsenet.org/journal/index.php/nct/article/view/0/45486
49. P.S. Sapaty, Spatial Grasp Model for Management of Dynamic Distributed Systems, *Acta Scientific Computer Sciences* 3(9) 2021. https://www.actascientific.com/ASCS/pdf/ASCS-03-0170.pdf
50. P.S. Sapaty, Spatial Grasp Model for Dynamic Distributed Systems, *Mathematical Machines and Systems* 3 (2021). http://www.immsp.kiev.ua/publications/articles/2021/2021_3/03_21_Sapaty.pdf
51. P.S. Sapaty, Development of Space-based Distributed Systems under Spatial Grasp Technology, *Mathematical Machines and Systems* 4 (2021).

Chapter 7

Advanced space projects management under Spatial Grasp Technology (SGT)

7.1 INTRODUCTION

A number of existing LEO space activities and related projects were reviewed in Chapter 2, also described in [1–23]. A number of developing space projects often relate to global defense like Strategic Defense Initiative (SDI) in the past [6–10] with multiple mini-satellites called "brilliant pebbles," and the recently launched notional architecture by Space Development Agency (SDA) [11–23]. The SDI was a long-term technology research program developed to examine the feasibility of developing defenses against a ballistic missile attack. The SDA plans to fight growing space-based threats, move quickly on hypersonic defense and track hypersonic threats from space, arm satellites with lasers to shoot down missiles, and so on. Unlike the SDI project, SDA architecture is oriented on intensive cooperation and collective behavior of many satellites and is made up of several layers, as follows (see more details in Chapter 2).

- *A space transport layer* is a global mesh network providing data and communications.
- *A tracking layer* provides tracking, targeting, and advanced warning of missile threats.
- *A custody layer* provides all-weather custody of identified time-critical targets.
- *A navigation layer* provides alternative positioning, navigation, and timing services.
- *A battle management layer* is a command, control, and communication network.
- *A support layer* includes ground command, control facilities, and user terminals.
- *A deterrence layer* will incubate new mission concepts, including in deep space.

We will be using here the developed Spatial Grasp model and Technology discussed in Chapters 3 and 4 and many existing publications on its previous

DOI: 10.1201/9781003230090-7

and current versions [24–53], including those for management of multiple satellite projects and architectures [46–53]. The exemplary solutions discussed below are mostly related to the SDI brilliant pebbles and SDA transport, tracking, and custody layers.

The rest of the chapter is organized as follows. *Section 7.2* reminds a brief summary of SGT described in detail in Chapters 3 and 4. *Section 7.3* gives an example of tasking brilliant pebbles in SGL. *Section 7.4* shows how self-spreading SGL scenarios can effectively track and destroy complexly moving objects (like hypersonic gliders) in large distributed spaces, and how to organize multithreaded tracing. *Section 7.5* demonstrates how to use mobile SGL code for implementation of custody layer for constant monitoring of terrestrial objects which may be of particular interest. *Section 7.6* shows how to organize cooperative work of custody and tracking layers under stable constellation topologies. *Section 7.7* concludes the chapter.

7.2 A BRIEF SUMMARY OF SPATIAL GRASP TECHNOLOGY

Within Spatial Grasp Technology (SGT), described in Chapters 3 and 4, also in more detail in [25, 30–35], a high-level scenario for any task to be performed in a distributed world is represented as an *active self-evolving pattern* rather than a traditional program, sequential or parallel one. This pattern, written in a recursive high-level Spatial Grasp Language (SGL) and expressing top semantics of the problem to be solved, can start from any point of the world. Then it spatially *propagates, replicates, modifies, covers, and matches* the distributed world in a parallel wavelike mode, while echoing the reached control states and data found or obtained for making decisions at higher levels and further space navigation, as symbolically shown in Figure 7.1.

Many spatial processes in SGL can start any time and in any place, cooperating or competing with each other, depending on applications. The self-spreading and self-matching SGL patterns-scenarios can create *active spatial infrastructures* covering any regions. These infrastructures can effectively support or express distributed knowledge bases, advanced command and control, situation awareness, autonomous and collective decisions, as well as any existing or hypothetical computational and/or control models,

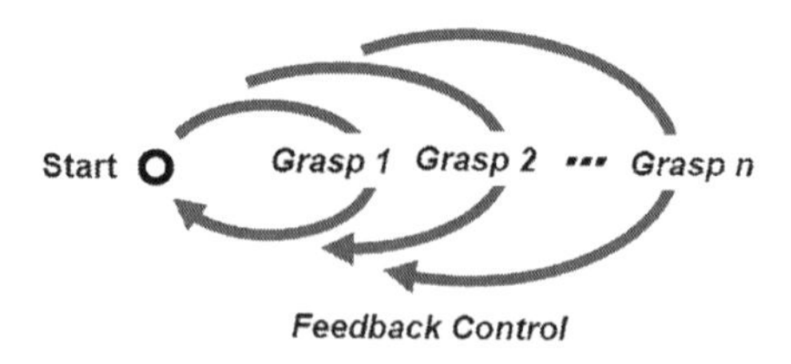

Figure 7.1 Controlled wavelike coverage and conquest of distributed spaces by SGT.

systems, and solutions. SGL has a deep recursive structure with its parallel spatial scenarios called *grasps* and universal operational and control constructs as *rules* (braces identifying repetition):

```
grasp → constant | variable | rule ({ grasp,})
```

7.3 TASKING BRILLIANT PEBBLES

We are starting here with Brilliant Pebbles, described in Chapter 2, also [6–10], which were designed to destroy ballistic missiles when they were most vulnerable, i.e., at first stages of their flight while still keeping many warheads to be released only later in the flight (see Figure 7.2).

The following functionality, loaded in all pebbles, is oriented on predominantly "boost" phase, assuming that the base stations put them on proper orbits where they will be staying unless discover and start pursuing the launched missiles.

```
hop(all_pebbles); nodal(Target);
repeat(
  Target = search(missile_launch);
  if(nonempty(Target), move_destroy(Target));
  sleep(delay))
```

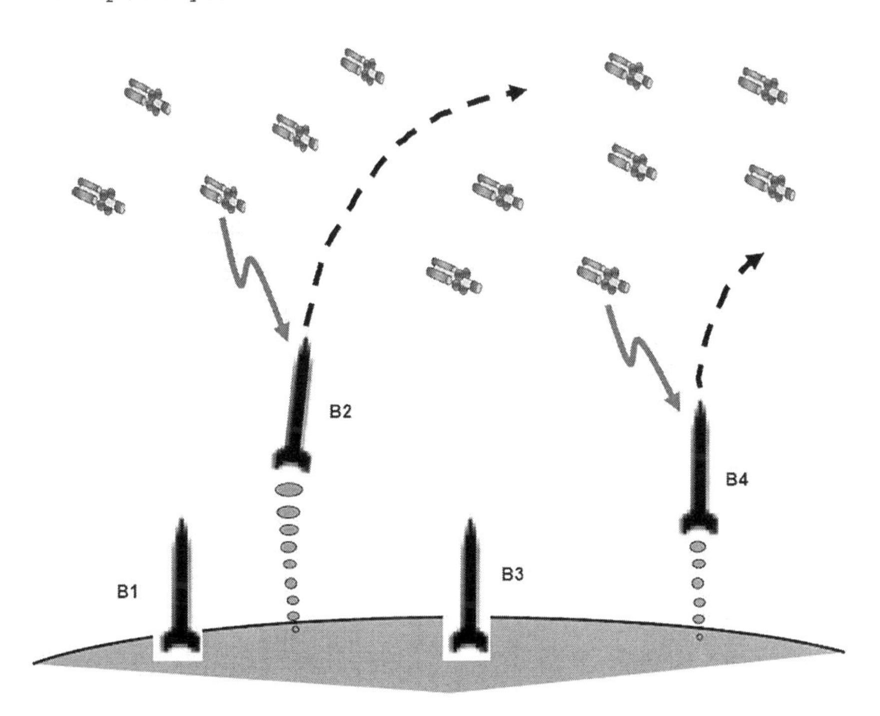

Figure 7.2. Brilliant pebbles and their operation.

Simulation of launching ballistic missiles B2 and B4 can be just as:

```
hop(B2, B4); activate(CONTENT)
```

Extended operation of pebbles for "boost," "post-boost," and possibly even "mid-course" stages of flight may be organized as follows, where pebbles begin continuous movement toward noticed missiles which may happen, at the beginning, to be far away from them. Pebbles will be making regular updates of the missile coordinates and try to reduce distance between them, each time at least on Threshold1 value. If pebbles appear closer to the missile on Threshold2 they will be trying to attack and finally destroy the missile in a collision impact.

```
hop(all_pebbles); nodal(Target);
nodal(Limit1 = ..., Limit2 = ...);
repeat(
   Target = search(missile_launch);
  if(nonempty(Target),
     repeat(
       update(Target); reduce(WHERE, Target, Limit1);
       if(close(WHERE, Target, Limit2), engage(Target))));
  sleep(delay))
```

Under this project, the pebbles were expected to behave mostly individually in pursuit of the missiles discovered, without or with minimum communication with other pebbles within their functionality. But in the SDA project, intensive communications and collective behavior of satellites are becoming crucial for solving different problems in space and on Earth, as explained in the subsequent sections.

7.4 MANAGING TRACKING LAYER

SDA's tracking layer [15, 18, 19, 22] will provide global indications, warning, tracking, and targeting of advanced missile threats, including hypersonic missile systems. This capability encompasses space-based sensing, as well as algorithms, novel processing schemes, data fusion across sensors and orbital regimes, and tactical data products able to be delivered to the appropriate user. The satellite constellation would be able to detect a hypersonic missile and to communicate among satellites and with the air and ground.

7.4.1 Tracing hypersonic gliders

Hypersonic weapons are breaking all the rules of traditional missile defense, as they are much harder to detect than the traditional ballistic missiles. Hypersonic weapons are maneuverable, meaning they can evade

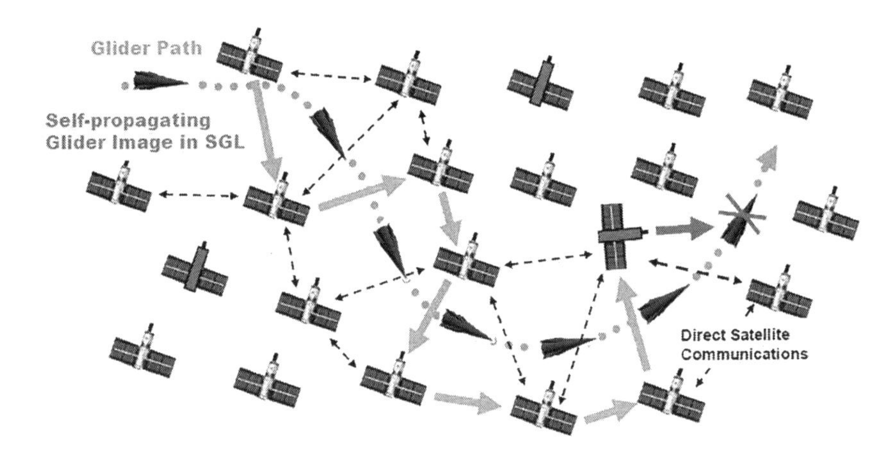

Figure 7.3 Tracing and destruction of hypersonic gliders by cooperating satellites.

ground-based sensors as they traverse the globe toward their target. With speed surpassing Mach 7 and the ability to maneuver mid-flight, they are potentially making the current defenses obsolete. The advanced sensors mounted on satellites can detect the threat and then regularly pass its details to neighboring LEO sensors with a better vision of the object. If to supply at least some satellites with a capability of impacting the moving gliders, they could be destroyed much more efficiently than from ground stations.

We can effectively organize large satellite networks to trace the hypersonic missiles wherever they go, where self-propagating their images in SGL will be keeping full control over them, as shown in Figure 7.3, and by the scenario that follows. Where `Threshold1` is indicating acceptable object's visibility from a satellite, and `Threshold2` is a minimal distance from which it may be attacked by a satellite that possesses such capability. At the beginning, all satellites are tasked with this scenario, and each begins continually looking for new moving objects, which when discovered are individually traced by the self-propagating spatial intelligence.

```
frontal(
 Depth = …, Type = …, Object, History, Threshold1 = …,
 Threshold2 = …;
 All = {stay, (hopfirst(Depth, Type); run(All))};
hopfirst(any, Type); run(All);
repeat(
  sleep(delay);
  Object = search(glider, new);
  visibility(Object) >= Threshold1;
  free_repeat(
    loop(
      visibility(Object) >= Threshold1;
```

```
      update(History, Object);
      if(and(danger(History), TYPE = destroyer,
             distance(Object, WHERE) < Threshold2),
         pursue_destroy(Object)));
   max_destination(hop(Depth, Type); visibility(Object));
   visibility(Object) >= Threshold1))
```

The moment when the satellite sensor sees a new object for the first time (within a given visibility threshold) is the start of a distributed tracing operation, and the object is continually monitored by this satellite until its visibility remains acceptable. Otherwise, the monitoring shifts to the neighbor which has the best vision of the object after analyzing its visibility in all neighbors. The history of the object movement and behavior can be collected and updated by each passed satellite by a SGT-produced mobile spatial intelligence individually assigned to this object and accompanying its physical movement via the satellite network. Depending on the collected history, it may be decided to destroy the object if the current satellite overseeing it has appropriate elimination capability. It may also be finally lost after passing through the whole satellites-controlled area.

7.4.2 Multithreaded tracing

Each such mobile intelligence is itself splitting into parallel branches entering all neighboring satellites to find the one with the best target visibility, into which the tracking scenario subsequently moves, and so on. So in this scenario, and as in Figure 7.3, it is supposed that tracing gliders by the satellite network is taking place by the resulting single mobile intelligence thread, where transference is always taking place to the satellite with best vision of the object. But in highly dynamic satellite network topologies, we can also

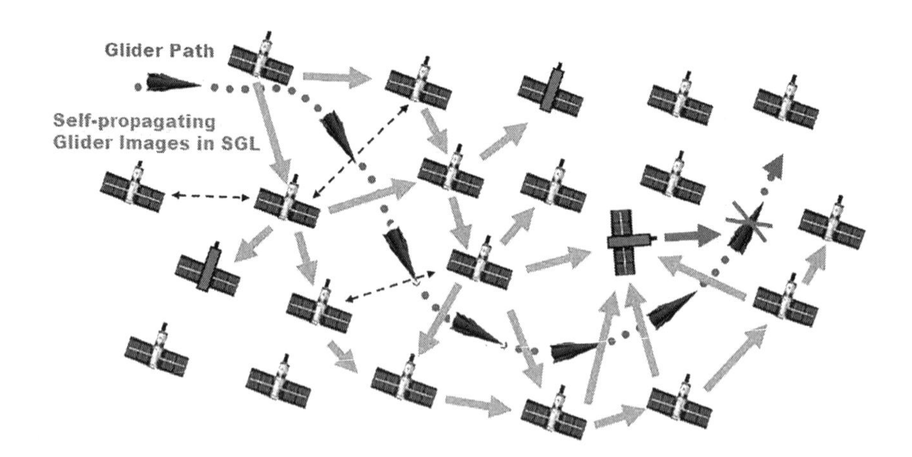

Figure 7.4 Multithread tracing example.

allow multithread tracing, where next tracing stages may take place from all neighbors with sufficient vision of the object. And even if the object is not currently visible from any neighbor after being lost by the current satellite, further flexibility can be introduced, where the tracing activity is transferred to some or all neighbors with any visibility, in hope to eventually reach satellites seeing this object, which, however, may not guarantee the final success too. The resulting SGL with the options mentioned may be as follows.

```
 frontal(
  Depth = …, Type = …, Object, History, Threshold1 = …,
  Threshold2 = …,
  All = {stay, (hopfirst(Depth, Type); run(All))};
hopfirst(any, Type); run(All);
repeat(
  sleep(delay);
  Object = search(glider, new);
  visibility(Object) >= threshold1;
  free_repeat(
    loop(
      visibility(Object) >= threshold1);
      update(History, Object);
      if(and(danger(History), TYPE = destroyer,
             distance(Object, WHERE) < threshold2),
         done(pursue_destroy(Object))));
    or_sequence(
      (hop(Depth, Type); visibility(Object) >= Threshold),
      hopforth(Depth, Type))))
```

The resultant mobile intelligence will be trying to move in between and via satellites unless the object becomes visible, but without returning to the previous satellites to block immediate cycling. The deeper cycles with retuning to previous satellites can appear, but may even be useful as the satellites move in space themselves and may change neighbors of each other, so the return may help find suitable neighbors this time, which can see the object. However, additional measures may be included into the scenario for not doing this endlessly if the object cannot be found at all. With these latest updates, the multithreaded tracking scenario may look like in Figure 7.4, but the satellite network topology can itself change during the object movement, so the shown communications may relate to quite different topologies changing in time.

7.5 MANAGING CUSTODY LAYER

The SDA Custody Layer [18, 19, 23] is oriented on all-weather custody of time-critical and mostly stationary targets on the Earth by Low Earth Orbit (LEO) satellites in orbits, which need to pass the observation duty to

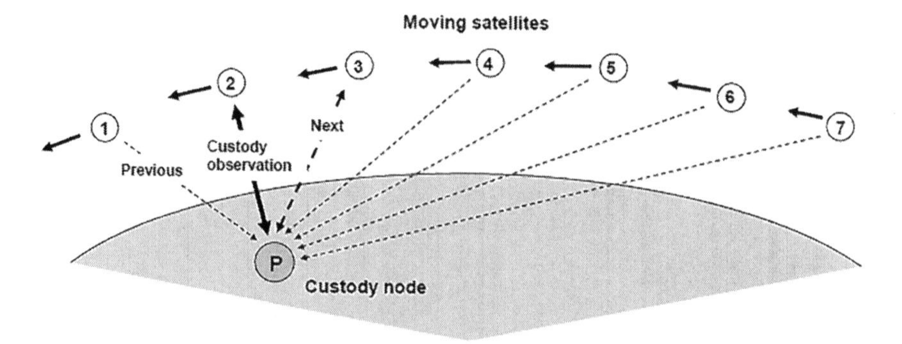

Figure 7.5 Observation of a custody node by moving satellites.

other satellites when moving away from the controlled objects, as shown in Figure 7.5. (For simplicity, only a single orbit with multiple satellites following each other in space is shown, whereas real satellite constellations may include many, as for the previous examples of tracing hypersonic objects, also with intersecting orbits.)

7.5.1 Continuous observation by a mobile scenario

Observation and monitoring of a stationary object on the Earth by moving LEO satellites may be, in some sense, considered as similar to the previous example where mobile objects were physically moving through the network of communicating satellites, which themselves were moving in space. The custody node may itself be considered as an object moving through the satellite network, and this can be effectively expressed in SGL. The movement from satellite to satellite will take place when a neighboring satellite appears to be closer to the same location on the Earth, into which the mobile intelligence with an updated history of the watched custody node will now move, and so on. This can be described by the following SGL scenario, which is also depicted in Figure 7.6.

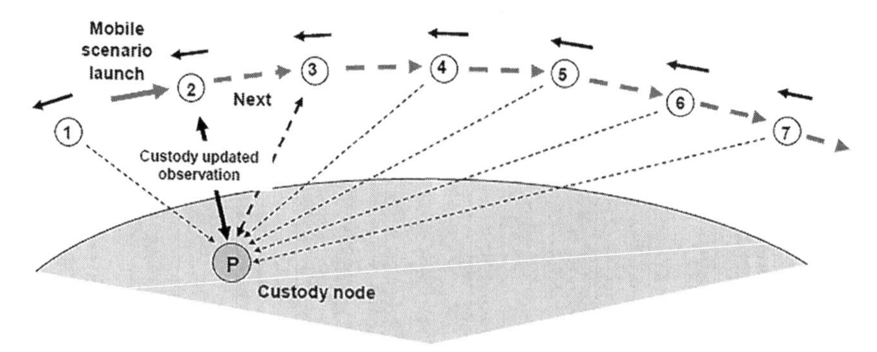

Figure 7.6 Managing custody observation by a mobile scenario in SGL.

```
frontal(Custody = X_Y, History, Depth = …, Type = …,
        Threshold = …);
hopfirst(any, Type);
repeat(distance(WHERE, Custody) > Threshold;
       hopfirst(Depth, Type));
repeat(
  if(distance(WHERE, Custody) <= Threshold,
     (update(History, observe(Custody));
      output(History); sleep(delay)),
     (min_destination(
       hop(Depth, Type); distance(WHERE, Custody));
      distance(WHERE, Custody) <= Threshold)))
```

Under the `output`, it meant that the system coordinating the custody service is being regularly informed, with details of access to it not detailed above. This system, which may be distributed itself, can also be part of the space constellation, with current satellite engaged in it too.

7.5.2 Working with constellation unpredictability

We may also make this scenario working with higher constellation unpredictability and dynamics, and in a multithreaded way, with more than a single satellite overseeing this custody at any time, similar to tracing hypersonic objects as before, with modification as follows.

```
frontal(Custody = X_Y, History, Depth = …, Type = …,
        Threshold);
hop(any, Type);
repeat(distance(WHERE, Custody) > Threshold;
       hopfirst(Depth, Type));
repeat(
  if(distance(WHERE, Custody) <= Threshold,
     (update(History, observe(Custody));
      output(History); sleep(delay)),
     or_sequence(
      (hop(Depth, Type);
       distance(WHERE, Custody) <= Threshold),
      hopforth(Depth, Type))))
```

7.6 WORKING WITH STABLE CONSTELLATION TOPOLOGIES

7.6.1 Managing custody alone

The previous scenarios can be simpler if satellite network is stable and well-structured, as follows, where `neighbors` mean nodes are directly linked to the current one in the network.

```
frontal(Custody = X_Y, History, Threshold = …);
min_destination(hop(all); distance(WHERE, Custody));
repeat(
  if(distance(WHERE, Custody) <= Threshold,
     (update(History, observe(Custody));
      output(History); sleep(delay)),
     min_destination(
       hop(neighbors); distance(WHERE, Custody))))
```

7.6.2 Custody and tracking working together

The following scenario can effectively integrate custody and tracking layers during Earth locations observation, discovery of launches, and tracing of complexly moving objects in LEO orbits by multiple satellites.

```
frontal(Custody = X_Y, History, Threshold1 = …,
        Threshold2 = …, Threshold3 = …);
min_destination(hop(all); distance(WHERE, Custody));
repeat(
  if(distance(WHERE, Custody) <= Threshold1,
     (update(History, observe(Custody);
      if(belong(glider_launch, History),
         free_repeat(
           loop(if(visibility(glider) >= Threshold2,
                   if(and(TYPE = destroyer,
                          distance(Object, WHERE) <
Threshold3),
                      pursue_destroy(glider))));
           max_destination(
               hop(neighbors); visibility(glider)))),
     min_destination(
       hop(neighbors); distance(WHERE, Custody))))
```

After fixing a launch at the custody object, the glider-tracing mobile intelligence is activated in SGL, which will accompany this glider wherever it goes via the satellite network. The SGL scenario will work repeatedly and endlessly, and if a new glider launch is detected in the observed custody, another tracing intelligence will be associated with this object and will follow it via the satellite networks, and so on. And at the same time the custody is updated too. Such integration of custody and tracking functionality is symbolically shown in Figure 7.7.

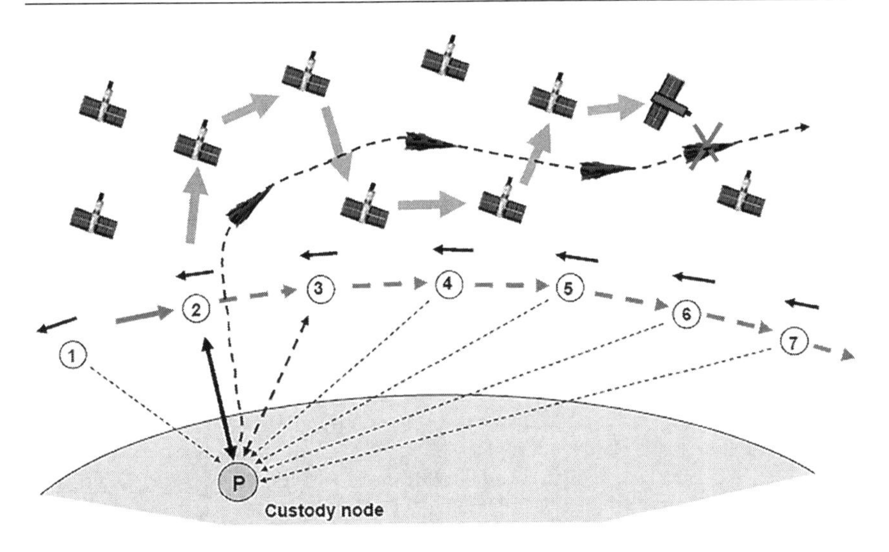

Figure 7.7 Cooperative work of custody and tracking layers.

7.7 CONCLUSIONS

The SGT self-evolving holistic parallel and distributed philosophy, model, and technology were found to be effectively matching the current and future plans of the space conquest, especially such recent project as the SDA New Space Architecture based on numerous cooperating satellites in predominantly LEO orbits. The main problem with cheap and relatively simple LEO satellites is that they are rapidly changing their positions over Earth locations, and to provide continuous observation, they regularly need to transfer their duties and accumulated information to other satellites. That can be effectively solved by the super-virus-like SGT and its basic SGL language, which navigate in parallel multi-satellite constellations and solve both continuous custody observation and tracing hypersonic gliders by collectively operating satellites. SGL spatial scenario mobility appeared to be extremely suitable to model, trace, and control complexly moving objects in space like hypersonic gliders, where this virtual to physical spatial mobility match can provide the unique and so far simplest solution for using multiple cooperating satellites for defense projects around the Earth. And for the continuous observation of custody objects on Earth by rapidly moving satellites, the chapter offered a solution where a custody node is treated itself as a mobile object moving through satellite constellations, elements of which are moving too. This unified "moving through" philosophy allowed us to effectively integrate custody and tracking functionalities into a unique solution for continuous observing of launches on Earth and subsequent immediate tracing of the launched objects wherever they can go.

REFERENCES

1. United Nations Register of Objects Launched into Outer Space, The United Nations Office for Outer Space Affairs. http://www.unoosa.org/oosa/en/spaceobjectregister/index.html
2. G. Martin, NewSpace: The «Emerging» Commercial Space Industry. https://ntrs.nasa.gov/archive/nasa/casi.ntrs.nasa.gov/20140011156.pdf
3. J.-M. Bockel, The Future of the Space Industry. *General Report*. 17 November, 2018. https://www.nato-pa.int/download-file?filename=sites/default/files/2018-12/2018%20-%20THE%20FUTURE%20OF%20SPACE%20INDUSTRY%20-%20BOCKEL%20REPORT%20-%20173%20ESC%2018%20E%20fin.pdf
4. M. Read, Space robotics market to reach $3.5bn by 2025: GMI report, 2019. https://satelliteprome.com/news/space-robotics-market-to-reach-3-5bn-by-2025-reveals-gmi-report/
5. Research on space debris, safety of space objects with nuclear power sources on board and problems relating to their collision with space debris. Committee on the Peaceful Uses of Outer Space. Vienna, 2019. http://www.unoosa.org/res/oosadoc/data/documents/2019/aac_105c_12019crp/aac_105c_12019crp_7_0_html/AC105_C1_2019_CRP07E.pdf
6. Strategic Defense Initiative. The White House. 1984. https://fas.org/irp/offdocs/nsdd/nsdd-119.pdf
7. The Strategic Defense Initiative: Program Facts. 1987. https://www.everycrsreport.com/files/19870722_IB85170_64d13e614c37eecbed39c00741ddfb269f814fef.pdf
8. J. Gattuso, Brilliant Pebbles: The Revolutionary Idea for Strategic Defense. 1990. https://www.heritage.org/defense/report/brilliant-pebbles-the-revolutionary-idea-strategic-defense.
9. "Brilliant Pebbles": The Revolutionary Idea for Strategic Defense. *The Heritage Foundation*, January 25, 1990. http://s3.amazonaws.com/thf_media/1990/pdf/bg748.pdf
10. Brilliant Pebbles. https://en.wikipedia.org/wiki/Brilliant_Pebbles.
11. S. Erwin, Space Development Agency could select three manufacturers to produce its next batch of satellites, *Space News*, April 14, 2021. https://spacenews.com/space-development-agency-could-select-three-manufacturers-to-produce-its-next-batch-of-satellites/
12. Broad Agency Announcement (BAA) National Defense Space Architecture (NDSA) Systems, Technologies, and Emerging Capabilities (STEC) Space Development Agency (SDA) HQ085021S0002. https://govtribe.com/file/government-file/hq085021s0002-sda-stec-dot-pdf
13. R.H. Freeman, Notional Satellite Architectures of Military (GEO-Earth) Versus Space Exploration (GEO-Mars), *MilsatMagazine*. 2020. http://www.milsatmagazine.com/story.php?number=500837950

14. Hypersonic Missile Defense: Issues for Congress, Congressional Research Service, 2021. https://fas.org/sgp/crs/weapons/IF11623.pdf
15. Miller, SDA Outlines Missile Tracking Satellite Plan, SDA, 2021. https://www.sda.mil/sda-outlines-missile-tracking-satellite-plan/
16. V. Machi, US Military Places a Bet on LEO for Space Security, *SDA*, 2021. https://www.sda.mil/us-military-places-a-bet-on-leo-for-space-security/
17. N. Strout, The Pentagon's new space agency has an idea about the future, July 3, 2019. https://www.c4isrnet.com/battlefield-tech/2019/07/03/the-pentagons-new-space-agency-has-an-idea-about-the-future/
18. Space Development Agency Next-Generation Space Architecture. 2019. https://www.airforcemag.com/PDF/DocumentFile/Documents/2019/SDA_Next_Generation_Space_Architecture_RFI%20(1).pdf
19. S. Magnuson, Web Exclusive: Details of the Pentagon's New Space Architecture Revealed. 2019. https://www.nationaldefensemagazine.org/articles/2019/9/19/details-of-the-pentagon-new-space-architecture-revealed
20. D. Messier, Space Development Agency Seeks Next-Gen Architecture in First RFI. 2019. http://www.parabolicarc.com/2019/07/07/space-development-agency-issues-rfi/
21. V. Insinna, Space agency has an ambitious plan to launch 'hundreds' of small satellites. Can it get off the ground? *DefenceNews Space*. 10 April, 2019. https://www.defensenews.com/space/2019/04/10/sda-has-an-ambitious-plan-to-launch-hundreds-of-small-satellites-can-it-get-off-the-ground/
22. Tracking, SDA, https://www.sda.mil/tracking/
23. Custody, SDA, https://www.sda.mil/custody/
24. P.S. Sapaty, WAVE-1: A new ideology of parallel processing on graphs and networks. *Proceedings of International Conference Frontiers in Computing*. Amsterdam, 1987. 10 p.
25. P.S. Sapaty, A distributed processing system: European Patent N 0389655. Publ. 10.11.93. 40 p.
26. P.S. Sapaty, M.J. Corbin, P.M. Borst, Mobile WAVE programming as a basis for distributed simulation and control of dynamic open systems. *Report at the 4th UK SIWG National Meeting, SGI Reality Centre, Theale, Reading*, October 11, 1994, 12 p.
27. P.S. Sapaty, M.J. Corbin, S. Seidensticker, Mobile intelligence in distributed Simulations. *Proceedings of the 14th Workshop on Standards for the Interoperability of Distributed Simulations* (IST UCF, Orlando, FL, March 1995). Orlando, 1995, 1045–1058.
28. P.S. Sapaty, P.M. Borst, M.J. Corbin, J. Darling, Towards the intelligent infrastructures for distributed federations. *Proceedings of 13th Workshop on Standards for the Interoperability of Distributed Simulations* (Orlando, FL, September 1995). Orlando, 1995, 351–366.
29. P.S. Sapaty, Mosaic Warfare: From Philosophy to Model to Solution, *Mathematical Machines and Systems* 3 (2019). http://www.immsp.kiev.ua/publications/articles/2019/2019_3/03_Sapaty_19.pdf
30. P.S. Sapaty, Symbiosis of Real and Simulated Worlds under Spatial Grasp Technology. Springer, 2021. 305 p.
31. P.S. Sapaty, Complexity in International Security: A Holistic Spatial Approach. Emerald Publishing, 2019. 160 p.
32. P.S. Sapaty, Holistic Analysis and Management of Distributed Social Systems. Springer, 2018. 234 p.

33. P.S. Sapaty, Managing Distributed Dynamic Systems with Spatial Grasp Technology. Springer, 2017. 284 p.
34. P.S. Sapaty, Ruling Distributed Dynamic Worlds. New York: John Wiley & Sons, 2005. 255 p.
35. P.S. Sapaty, Mobile Processing in Distributed and Open Environments. New York: John Wiley & Sons, 1999. 410 p.
36. P.S. Sapaty, Global Network Management under Spatial Grasp Paradigm. *International Robotics & Automation Journal*, 6(3) 2020, 134–148. https://medcraveonline.com/IRATJ/IRATJ-06-00212.pdf
37. P. S. Sapaty, Global Network Management under Spatial Grasp Paradigm, *Global Journal of Researches in Engineering: J General Engineering*, 20(5) (2020), 58–81. Version 1.0 Year 20. https://globaljournals.org/GJRE_Volume20/6-Global-Network-Management.pdf
38. P.S. Sapaty, Advanced terrestrial and celestial missions under spatial grasp technology, *Aeronautics and Aerospace Open Access Journal*, 4(3) (2020). https://medcraveonline.com/AAOAJ/AAOAJ-04-00110.pdf
39. P.S. Sapaty, Spatial Management of Distributed Social Systems, *Journal of Computer Science Research*, 2(3) (2020, July). https://ojs.bilpublishing.com/index.php/jcsr/article/view/2077/pdf
40. P.S. Sapaty, Towards Global Nanosystems under High-level Networking Technology, *Acta Scientific Computer Sciences*, 2(8) (2020). https://www.actascientific.com/ASCS/pdf/ASCS-02-0051.pdf
41. P.S. Sapaty, Symbiosis of Distributed Simulation and Control under Spatial Grasp Technology, *SSRG International Journal of Mobile Computing and Application (IJMCA)*, 7(2) (2020, May–August). http://www.internationaljournalssrg.org/IJMCA/2020/Volume7-Issue2/IJMCA-V7I2P101.pdf
42. P.S. Sapaty, Global Network Management under Spatial Grasp Paradigm, *International Robotics & Automation Journal*, 6(3) (2020). https://medcraveonline.com/IRATJ/IRATJ-06-00212.pdf
43. P.S. Sapaty, Global Network Management under Spatial Grasp Paradigm, *Global Journal of Researches in Engineering: J General Engineering*, 20(5) (2020). https://globaljournals.org/GJRE_Volume20/6-Global-Network-Management.pdf
44. P.S. Sapaty, Symbiosis of Virtual and Physical Worlds under Spatial Grasp Technology, *Journal of Computer Science & Systems Biology*, 13(6) (2020). https://www.hilarispublisher.com/open-access/symbiosis-of-virtual-and-physical-worlds-under-spatial-grasp-technology.pdf
45. P.S. Sapaty, Symbiosis of Real and Simulated Worlds under Global Awareness and Consciousness, *The Science of Consciousness* | *TSC* 2020. https://eagle.sbs.arizona.edu/sc/report_poster_detail.php?abs=3696
46. P.S. Sapaty, Spatial Grasp as a Model for Space-based Control and Management Systems, *Mathematical Machines and Systems*, 1 (2021), 135–138. http://www.immsp.kiev.ua/publications/articles/2021/2021_1/Sapaty_book_1_2021.pdf
47. P.S. Sapaty, Managing multiple satellite architectures by spatial grasp technology, *Mathematical Machines and Systems*, 1 (2021), 3–16. http://www.immsp.kiev.ua/publications/eng/2021_1/
48. P.S. Sapaty, Spatial Management of Large Constellations of Small Satellites, *Mathematical Machines and Systems*, 2 (2021). http://www.immsp.kiev.ua/publications/articles/2021/2021_2/02_21_Sapaty.pdf

49. P.S. Sapaty, Global Management of Space Debris Removal Under Spatial Grasp Technology, *Acta Scientific Computer Sciences*, 3(7) (2021, July). https://www.actascientific.com/ASCS/pdf/ASCS-03-0135.pdf
50. P.S. Sapaty, Space Debris Removal under Spatial Grasp Technology, *Network and Communication Technologies*, 6(1) (2021). https://www.ccsenet.org/journal/index.php/nct/article/view/0/45486
51. P.S. Sapaty, Spatial Grasp Model for Management of Dynamic Distributed Systems, *Acta Scientific Computer Sciences*, 3(9) (2021). https://www.actascientific.com/ASCS/pdf/ASCS-03-0170.pdf
52. P.S. Sapaty, Spatial Grasp Model for Dynamic Distributed Systems, *Mathematical Machines and Systems*, 3 (2021). http://www.immsp.kiev.ua/publications/articles/2021/2021_3/03_21_Sapaty.pdf
53. P.S. Sapaty, Development of Space-based Distributed Systems under Spatial Grasp Technology, *Mathematical Machines and Systems*, 4 (2021).

Chapter 8

Using virtual layer for constellation management

8.1 INTRODUCTION

In Chapters 6 and 7, we discussed basic constellation management and its use for solving important practical problems [1–23], including those related to global security and defense [6–23]. This was with the use of Spatial Grasp Technology (SGT) and its basic high-level Spatial Grasp Language (SGL) described in detail in Chapters 3 and 4, development of which is covering a rather long history, with only some activities mentioned in [24–53]. This chapter goes further from the previous sections, actually higher, showing how to introduce an SGT-based virtual or overseeing layer over the distributed satellite constellations, and how this layer can simplify vision, formulation, and solving even more complex problems, both celestial and terrestrial, by collectively behaving satellites. The exemplary solutions described below mostly relate to the SDA Space Architecture discussed in Chapters 2 and 7, and also [11–23], and especially to its custody and battle management layers [22, 23].

The rest of the chapter is organized as follows. *Section 8.2* provides a brief summary of SGT described in detail in Chapters 3 and 4. *Section 8.3* introduces virtual layer for the custody support, where custody-related virtual nodes are constantly managed and updated regardless of satellites movement is space. *Section 8.4* provides examples of custody operations via the virtual layer, including verifying changing distances between remote custody locations, using regularly updated virtual layer for worldwide delivery of goods, and analyzing complex distributed system as a whole on example of forest fires covering different regions with different fire intensities. *Section 8.5* describes distributed virtual-physical command and control with regularly updating correspondence between virtual and physical (or earth-based) C2, and delivering C2 operations via virtual layer and managing two- and then arbitrary multilevel virtual-physical C2 Infrastructure. This may allow us to manage complex terrestrial, celestial. and combined missions even completely from space, if needed. *Section 8.6* concludes the chapter.

DOI: 10.1201/9781003230090-8

8.2 A BRIEF SUMMARY OF SPATIAL GRASP TECHNOLOGY

Within SGT, described in Chapters 3–4, also in more detail in [25, 30–35], a high-level scenario for any task to be performed in a distributed world is represented as an *active self-evolving pattern* rather than a traditional program, sequential or parallel one. This pattern, written in a recursive high-level SGL and expressing top semantics of the problem to be solved, can start from any point of the world. Then it spatially *propagates, replicates, modifies, covers, and matches* the distributed world in a parallel wavelike mode, while echoing the reached control states and data found or obtained for making decisions at higher levels and further space navigation, as symbolically shown in Figure 8.1.

Many spatial processes in SGL can start any time and in any place, cooperating or competing with each other, depending on applications. The self-spreading and self-matching SGL patterns-scenarios can create *active spatial infrastructures* covering any regions. These infrastructures can effectively support or express distributed knowledge bases, advanced command and control, situation awareness, autonomous and collective decisions, as well as any existing or hypothetical computational and/or control models, systems, and solutions. SGL has a deep recursive structure with its parallel spatial scenarios called *grasps* and universal operational and control constructs as *rules* (braces identifying repetition):

```
grasp  →  constant | variable | rule ({ grasp,})
```

8.3 INTRODUCING VIRTUAL LAYER FOR CUSTODY SUPPORT

The custody layer may need constant observation not only of an object or objects at a certain location, but also those distributed over a certain (even vast) area and connected with each other, like a spatial infrastructure. In addition, this may need involvement of a number of satellites simultaneously overseeing different parts of this infrastructure to work together as a system. Such organization can be effectively provided under SGT too,

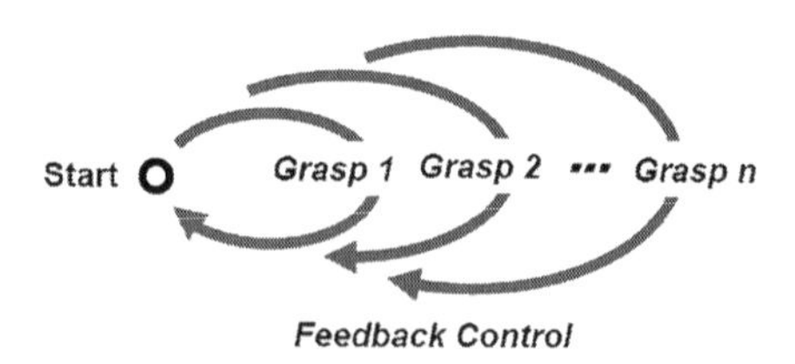

Figure 8.1 Controlled wavelike coverage and conquest of distributed spaces by SGT.

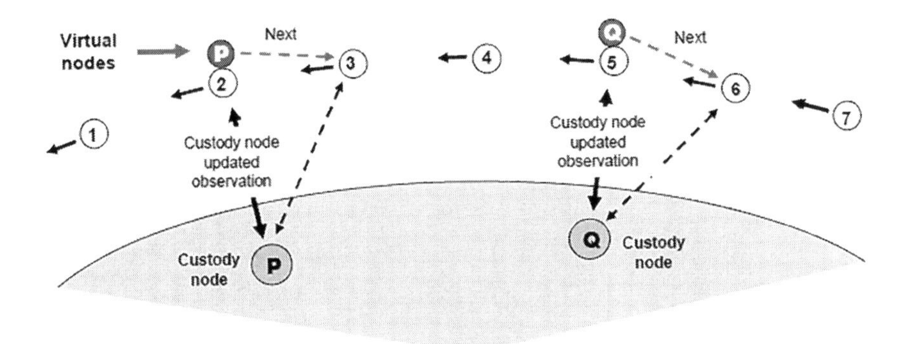

Figure 8.2 Introduction of virtual custody layer.

by introducing a higher, virtual, custody layer continually overseeing the corresponding physical objects and their infrastructures, regardless of constant movement of satellites around the globe. An example of such a layer consisting of two virtual nodes P and Q, whose physical nodes are similarly named, is presented in Figure 8.2. The following SGL scenario creates and constantly supports such virtual nodes located in substituting each other satellites.

```
frontal(Name, Custody, Threshold = …);
nodal(History);
parallel(
 (Name = P; Custody = XP_YP),
 (Name = Q; Custody = XQ_YQ));
min_destination(
  hop_sats(all); distance(WHERE, Custody));
create_node(Name);
repeat(
  if(distance(WHERE, Custody) <= Threshold,
     (update(Custody);
      update(History, observe(Custody)); sleep(delay)),
     DOER = min_destination(
      hop_sats(Neighbors); distance(WHERE, Custody))))
```

Using special environmental variables designated as DOER, the virtual P and Q nodes are regularly relocated to other neighboring satellites when the current ones, hosting these nodes, move away from the ground locations they have to observe. This constantly preserves spatial correspondence between physical nodes and their virtual copies regardless of the movement of satellites. (Of course, if there are enough of them effectively distributed throughout the globe, or at least are always present around the area of interest, and they can directly communicate with each other.)

Any semantic relations can be established between such virtual nodes which may reflect different links or connections between the related physical

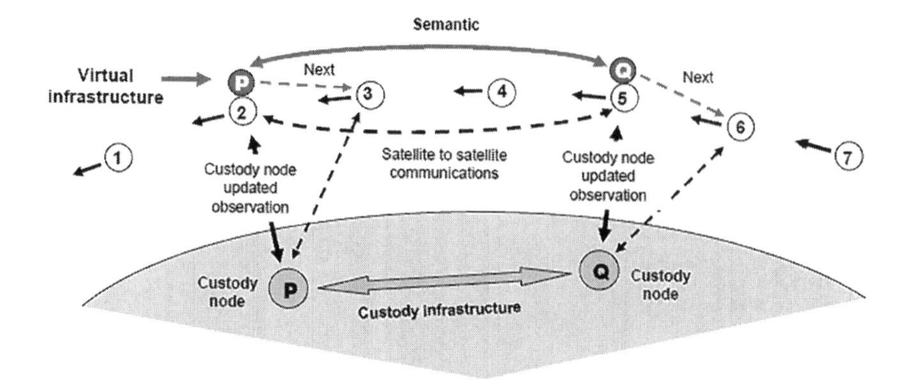

Figure 8.3 Establishing semantic link between virtual nodes.

custody nodes (for example, roads, communication channels, or command and control subordination), as shown in Figure 8.3, with a `Semantic` link set up between these nodes by:

```
hop_node(P); linkup(Semantic, node(Q)) or
hop_node(Q); linkup(Semantic, node(P))
```

The virtual nodes connected with each other by a semantic relation will always be preserving this relation between them on the internal SGL interpretation level, which automatically adjusts to the propagation of nodes between moving satellites, the latter being their temporary hosts or "doers."

Such virtual layer with any number of nodes and any number and types of semantic links between them can be effectively supported under SGT in the constellations of moving and communicating satellites, with examples of some elementary operations using this layer shown in the subsequent sections.

If semantic links between virtual nodes reflect physical links between custody nodes, and the physical distance between these custody nodes allows communication between satellites which happened to be over them, then traversing of such semantic links can be implemented by direct contacts between the satellites or via shortest paths between them in case of stable network topology.

8.4 EXAMPLES OF CUSTODY OPERATIONS VIA THE VIRTUAL LAYER

8.4.1 Verifying changing distance between remote custody locations

Directly starting from virtual node `Q`, the following scenario lifts the related current custody physical coordinates, goes with them to the virtual node `P` via the semantic links between them, and lifts their related physical custody coordinates as well, see also Figure 8.3. Then it calculates physical distance between the two physical custody nodes, and if it finds out that the nodes are too close to (or on the opposite, far away from) each other, issues a «danger» message with the distance found, which will be delivered to some global management level.

```
Nodal(Custody);
frontal(Cust, Threshold = distance, Distance);
hop_node(Q);
whirl(
  Cust = Custody; hop_link(Semantic);
  Distance = distance(Cust, Custody);
  if(Distance <= Threshold,
     output("Danger", Distance));
  sleep(delay))
```

8.4.2 Organizing worldwide goods delivery via virtual layer

We will show here how the use of space virtual layer describing possible routes between custody nodes may help to organize terrestrial goods delivery from a source to destination, with virtual links holding physical distances between the related physical locations, as shown in Figure 8.4.

Let us show how to organize in SGL a goods delivery from the node 3 to the destination 19, where the spatial scenario starts in virtual node 3, finds shortest path to node 19, as in Figure 8.5, collects and returns coordinates

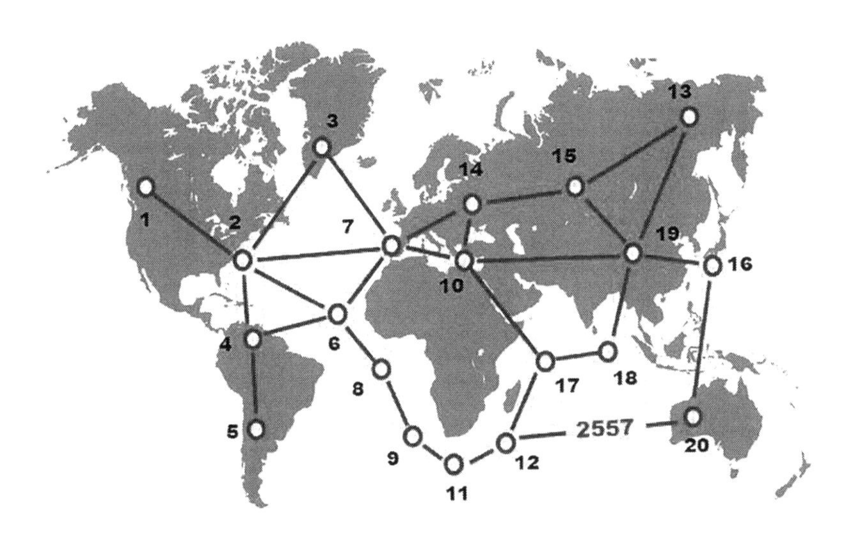

Figure 8.4 Space-based virtual layer for goods delivery on Earth.

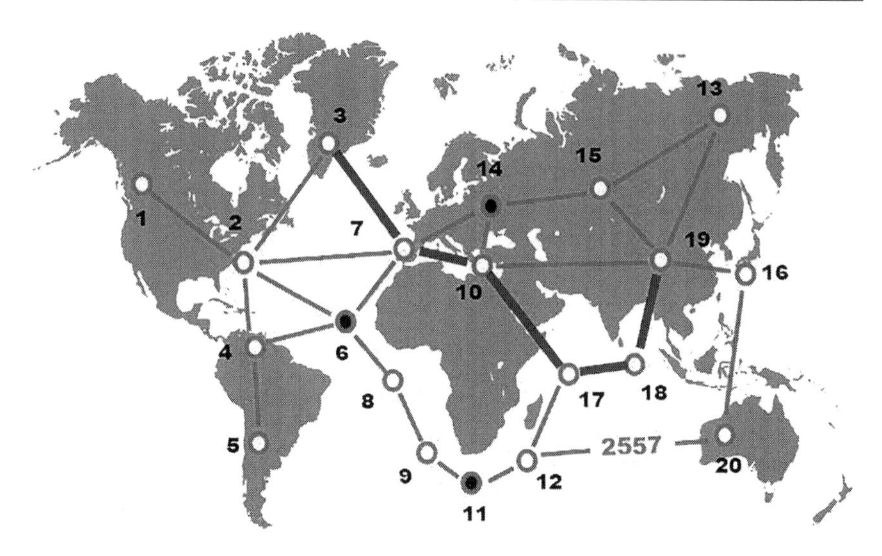

Figure 8.5 Terrestrial goods delivery via shortest path found on virtual layer.

of physical locations reflecting this path to node 3, and then passes them to the terrestrial layer for physical delivery.

```
frontal(Far, Dest = 19); nodal(Distance, Up);
sequence(
 (hop_node(3); Distance = 0;
  repeat(
    hop_links(all); CONTENT == "allowed";
    increment(Far, LINK);
    or(Distance == nil, Distance > Far);
    Distance = Far; Up = BACK)),
 (hop_node(19); frontal(Path);
  repeat(Path = append(custody(NAME), Path); hop(Up));
  record(Path)))
```

Stepwise movement of physical goods step by step by the found full path to the destination can be described as:

```
frontal(Goods = package, Path = recorded);
move_custody(3);
repeat(move_custody(withdraw(Path, 1)));
delivery_report(Goods)
```

In case of a dynamic physical world, where some virtual nodes (shown in black in Figure 8.5) may happen to reflect custody positions which cannot be used at that moment of time (say, influenced by disasters or conflicts), physical move to the chosen destination may need redefining shortest path to it from every new physical position reached. For such a case, we can organize the scenario above as a procedure `Pathfinding`, which will be activated for each currently achieved position in physical space, with using each time only the first item of the regularly updated path, as follows.

```
frontal(Goods = package, Path);
move_custody(3);
repeat(run(Pathfinding); move_custody(withdraw(Path, 1))
delivery_report(Goods)
```

8.4.3 Analyzing complex distributed system as a whole

a) *Creating virtual nodes relating to the areas of fire on Earth, as in Figure 8.6.*
   This can be done by the following scenario:

```
nodal(Name, Level);
parallel(
 (Name = 1; Custody = X1_Y1),
 (Name = 2; Custody = X2_Y2),
 (Name = 3; Custody = X3_Y3),
 (Name = 4; Custody = X4_Y4));
min_destination(
  hop_sats(all); distance(WHERE, Custody));
create_node(Name);
repeat(
  if(distance(WHERE, Custody) <= Threshold,
      (Level = (access(Custody); intensity(fire));
sleep(delay)),
      DOER = min_destination(
       hop_sats(neighbors); distance(WHERE, Custody))))
```

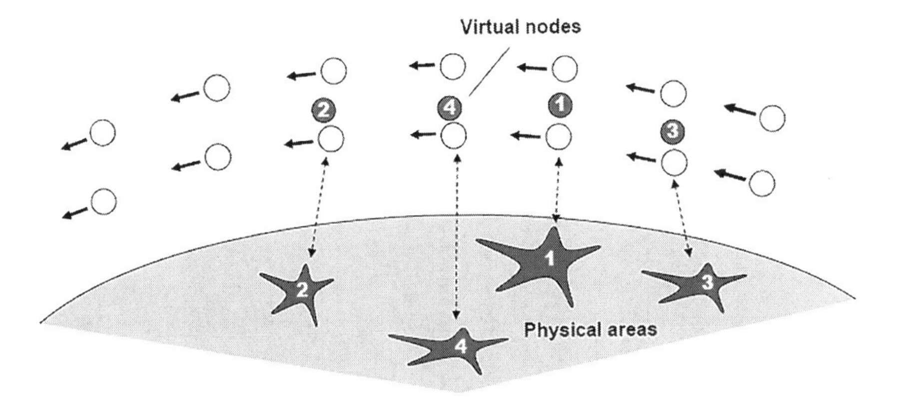

Figure 8.6 Setting up virtual nodes relating to areas of fire on Earth.

b) *Finding region name under the maximum fire:*

```
Max = max(hop_nodes(all); append(Level, NAME));
Region = Max[2]
```

The result for Figure 8.6 will be: `1`.

c) *Setting neighbor-type virtual links for regions that are physically close to each other:*

```
frontal(Cust, Name, Threshold = …);
hop_nodes(all); Name = NAME; Cust = Custody;
hop_nodes(all_other);
if(distance(Cust, Custody) < Threshold,
   if(Name > NAME, linkup(neighbor, BACK)))
```

See Figure 8.7 for all such links established.

d) *Finding the region with maximum fire and requesting all neighboring regions for assistance:*

```
frontal(Max, Area, Height);
Max = max(hop_nodes(all); append(Level, NAME));
hop_node(Max[2]); Area = Custody; Height = Level;
hop_links(neighbor);
access(Custody); assist(Area, Height, Level)
```

As the result, region `1` with maximum level of fire will be requesting neighboring regions `2`, `3`, and `4` for help.

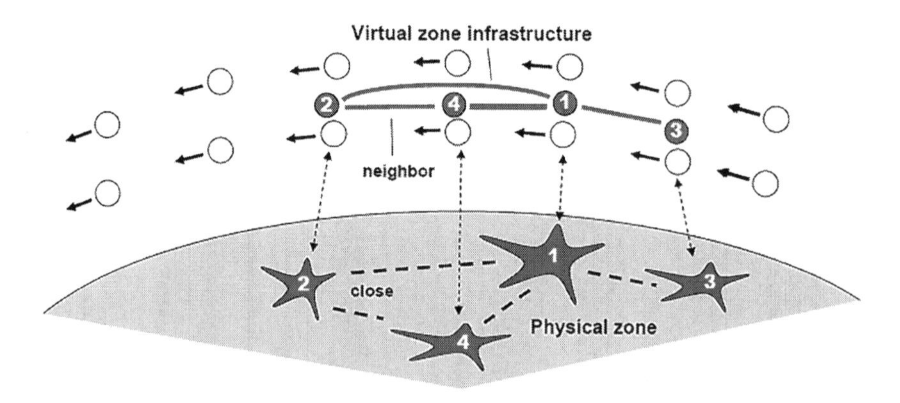

Figure 8.7 Setting neighbor-type links between virtual nodes.

8.5 DISTRIBUTED VIRTUAL-PHYSICAL COMMAND AND CONTROL (C2)

Establishing celestial virtual layer for complex distributed command and control systems on Earth can essentially improve vision, understanding, and management of large systems, defense ones including, especially in complex situations on Earth and difficulties of communications along large distances. The load of effective control of large operations and campaigns may shift to space, even completely if needed. We will consider here different interactions between terrestrial and celestial C2 versions.

8.5.1 Regularly updating correspondence between virtual and physical Command and Control (C2)

We are showing here how correspondence between terrestrial command and control location and its virtual celestial copy is being constantly updated in both ways.

```
frontal(Name = P, GrC2 = XP_YP, Threshold = …);
min_destination(
   hop_sats(all); distance(WHERE, GrC2));
create_node(Name);
repeat(
  update(History, access(GrC2)); sleep(delay));
  update(access(GrC2), History); sleep(delay));
  if(distance(WHERE, GrC2) > Threshold,
     DOER = min_destination(
              hop_sats(neighbors); distance(WHERE, GrC2))))
```

8.5.2 Delivering C2 operation via virtual layer

Delivering C2 operation starting from ground GrC2, and then from P (chief) to node Q (subordinate) virtual nodes interlinked by oriented control, and ultimately down to GrC2:

```
frontal(Order) = actions;
hop(P); update(History, Order);
hop(link(+control), node(Q));
update(History, Order);
access(GrC2); execute(Order)
```

8.5.3 Managing two-level virtual-physical infrastructure

Regular local observation and tasking of subordinates in the two-level hierarchy, with oriented top-down links control, and generating command

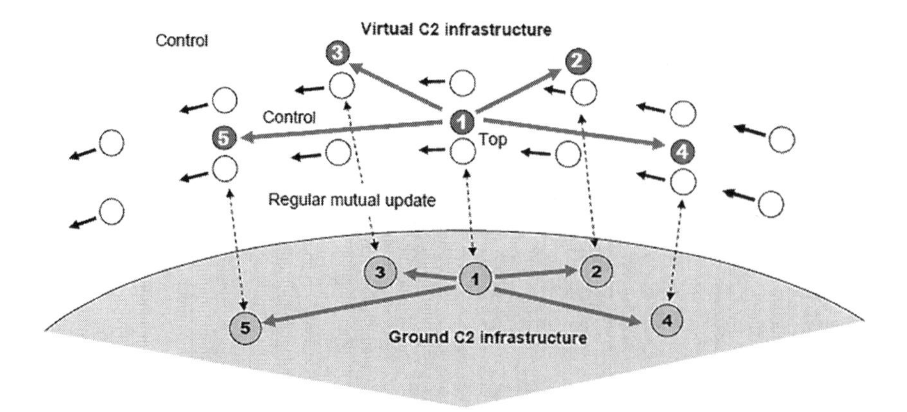

Figure 8.8 Two-level physical-virtual C2 infrastructure.

at the top to be executed at the bottom level (see also Figure 8.8) will be as follows.

```
nodal(History, Situation, Top = "1");
frontal (Command);
hop(Top);
whirl(
  Situation =
     analyse(History, (hop_links(+control); History));
  If(Situation == to_be_changed,
     (Command = react(Situation);
      update(History, Command);
      update(access(GrC2), Command);
      hop_links(+control); update(History, Command);
      access(GrC2); execute(Command)));
  sleep(delay))
```

8.5.4 Multilevel virtual-physical C2 infrastructure

Regular observation and tasking of the arbitrary multilevel hierarchical system, as in Figure 8.9, will be as follows.

```
nodal(History, Situation, Top = "1");
frontal(
  Command,
  Downup =
   {unit(append(
     History, (hop_links(+control); run(Downup))))};
hop(Top);
whirl(
  Situation = analyse(run(Downup));
  If(Situation == to_be_changed,
```

```
    (Command = react(Situation);
     repeat(
       update(Command, History);
       update(History, Command);
       stay(update(access(GrC2), Command);
            update(Command, access(GrC2)));
       hop_links(+control));
     access(GrC2); execute(Command));
sleep(delay))
```

This scenario is regularly analyzing the overall situation, making top decision and distributing it to all subordinate systems, with correction at each level. Using rule `stay` in case of failures for updates with the ground at different levels allows us not to stop the whole process in any case, which will continue to be managed from the virtual level.

8.6 CONCLUSIONS

It has been shown that by establishing a special virtual layer over constellations of satellites it becomes possible to implement higher-level functions and supervise any other layers. This may allow, if needed, for full control and management from space of any terrestrial and celestial operations, missions, and campaigns. It was demonstrated how to introduce this SGT-based virtual or overseeing layer and how this layer can simplify vision, formulation, and solving complex problems, both celestial and terrestrial, by collectively behaving satellites, where custody-related virtual nodes are constantly managed and updated regardless of satellites' fast movement in space and their short visibility from Earth locations. In exemplary applications, it was discussed how regularly updated virtual layer can be useful for

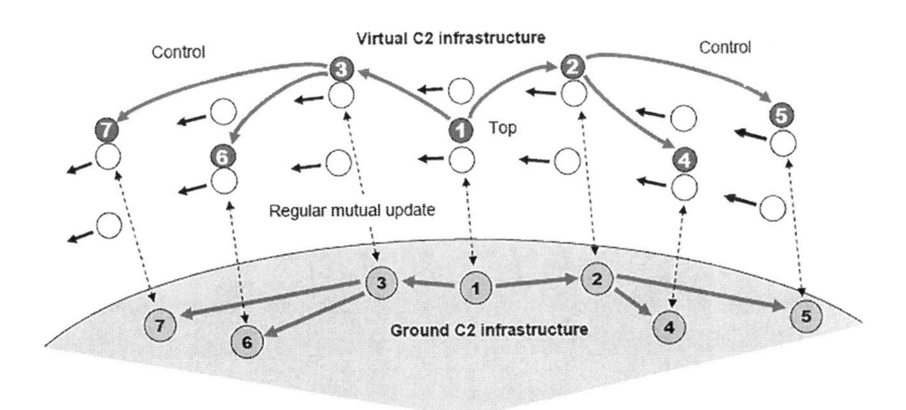

Figure 8.9 Multilevel physical-virtual C2 infrastructure.

worldwide terrestrial delivery of goods, overseeing corresponding physical objects and their infrastructures on Earth, as well as implementation of distributed virtual-physical command and control with effective management of multilevel virtual-physical C2 infrastructures. The described exemplary solutions can also be useful for the SDA Space Architecture (discussed in Chapters 2 and 7, also [11–23]), and especially for its custody and battle management layers [22, 23].

REFERENCES

1. United Nations Register of Objects Launched into Outer Space, The United Nations Office for Outer Space Affairs. http://www.unoosa.org/oosa/en/spaceobjectregister/index.html.
2. G. Martin, NewSpace: The «Emerging» Commercial Space Industry. https://ntrs.nasa.gov/archive/nasa/casi.ntrs.nasa.gov/20140011156.pdf
3. J.-M.Bockel, The Future of the Space Industry. *General Report*. 2018. Nov. 17. https://www.nato-pa.int/download-file?filename=sites/default/files/2018-12/2018%20-%20THE%20FUTURE%20OF%20SPACE%20INDUSTRY%20-%20BOCKEL%20REPORT%20-%20173%20ESC%2018%20E%20fin.pdf
4. M. Read, Space robotics market to reach $3.5bn by 2025: GMI report, 2019. https://satelliteprome.com/news/space-robotics-market-to-reach-3-5bn-by-2025-reveals-gmi-report/
5. Research on space debris, safety of space objects with nuclear power sources on board and problems relating to their collision with space debris. Committee on the Peaceful Uses of Outer Space. Vienna, 2019. http://www.unoosa.org/res/oosadoc/data/documents/2019/aac_105c_12019crp/aac_105c_12019crp_7_0_html/AC105_C1_2019_CRP07E.pdf
6. Strategic Defense Initiative. The White House. 1984. https://fas.org/irp/offdocs/nsdd/nsdd-119.pdf
7. The Strategic Defense Initiative: Program Facts. 1987. https://www.everycrsreport.com/files/19870722_IB85170_64d13e614c37eecbed39c00741ddfb269f814fef.pdf
8. Gattuso J. Brilliant Pebbles: The Revolutionary Idea for Strategic Defense. 1990. https://www.heritage.org/defense/report/brilliant-pebbles-the-revolutionary-idea-strategic-defense
9. "Brilliant Pebbles": The Revolutionary Idea for Strategic Defense. The Heritage Foundation, January 25, 1990. http://s3.amazonaws.com/thf_media/1990/pdf/bg748.pdf
10. Brilliant Pebbles. https://en.wikipedia.org/wiki/Brilliant_Pebbles.
11. S. Erwin, Space Development Agency could select three manufacturers to produce its next batch of satellites, Space News, April 14, 2021. https://spacenews.com/space-development-agency-could-select-three-manufacturers-to-produce-its-next-batch-of-satellites/
12. Broad Agency Announcement (BAA) National Defense Space Architecture (NDSA) Systems, Technologies, and Emerging Capabilities (STEC) Space Development Agency (SDA) HQ085021S0002. https://govtribe.com/file/government-file/hq085021s0002-sda-stec-dot-pdf

13. R.H. Freeman, Notional Satellite Architectures of Military (GEO-Earth) Versus Space Exploration (GEO-Mars), *MilsatMagazine*. 2020. http://www.milsatmagazine.com/story.php?number=500837950
14. Hypersonic Missile Defense: Issues for Congress, Congressional Research Service, 2021. https://fas.org/sgp/crs/weapons/IF11623.pdf
15. Miller, SDA Outlines Missile Tracking Satellite Plan, *SDA*, 2021. https://www.sda.mil/sda-outlines-missile-tracking-satellite-plan/
16. V. Machi, US Military Places a Bet on LEO for Space Security, *SDA*, 2021. https://www.sda.mil/us-military-places-a-bet-on-leo-for-space-security/
17. N. Strout, The Pentagon's new space agency has an idea about the future, July 3, 2019. https://www.c4isrnet.com/battlefield-tech/2019/07/03/the-pentagons-new-space-agency-has-an-idea-about-the-future/
18. Space Development Agency Next-Generation Space Architecture. 2019. https://www.airforcemag.com/PDF/DocumentFile/Documents/2019/SDA_Next_Generation_Space_Architecture_RFI%20(1).pdf
19. S. Magnuson, Web Exclusive: Details of the Pentagon's New Space Architecture Revealed. 2019. https://www.nationaldefensemagazine.org/articles/2019/9/19/details-of-the-pentagon-new-space-architecture-revealed
20. D. Messier, Space Development Agency Seeks Next-Gen Architecture in First RFI. 2019. http://www.parabolicarc.com/2019/07/07/space-development-agency-issues-rfi/
21. V. Insinna, Space agency has an ambitious plan to launch 'hundreds' of small satellites. Can it get off the ground? *DefenceNews*. Space. 10 April, 2019. https://www.defensenews.com/space/2019/04/10/sda-has-an-ambitious-plan-to-launch-hundreds-of-small-satellites-can-it-get-off-the-ground/
22. CUSTODY, SDA. https://www.sda.mil/custody/
23. Battle Management Command, Control, and Communication (BMC3), SDA. https://www.sda.mil/battle-management/
24. P.S. Sapaty, WAVE-1: A new ideology of parallel processing on graphs and networks. *Proceedings of the International Conference of Frontiers in Computing*. Amsterdam, 1987. 10 p.
25. P.S. Sapaty, A distributed processing system: European Patent N 0389655. Publ. 10.11.93. 40 p.
26. P.S. Sapaty, M.J. Corbin, P.M. Borst, Mobile WAVE programming as a basis for distributed simulation and control of dynamic open systems. *Report at the 4th UK SIWG National Meeting, SGI Reality Centre, Theale, Reading*, October 11, 1994, 12 p.
27. P.S. Sapaty, M.J. Corbin, S. Seidensticker, Mobile intelligence in distributed Simulations. *Proceedings of the 14th Workshop on Standards for the Interoperability of Distributed Simulations* (IST UCF, Orlando, FL, March 1995). Orlando, 1995, 1045–1058.
28. P.S. Sapaty, P.M. Borst, M.J. Corbin, J. Darling, Towards the intelligent infrastructures for distributed federations. *Proceedings of the 13th Workshop on Standards for the Interoperability of Distributed Simulations* (Orlando, FL, September 1995). Orlando, 1995, 351–366.
29. P.S. Sapaty, Mosaic Warfare: From Philosophy to Model to Solution, *Mathematical Machines and Systems*, 3 (2019). http://www.immsp.kiev.ua/publications/articles/2019/2019_3/03_Sapaty_19.pdf
30. P.S. Sapaty, Symbiosis of Real and Simulated Worlds under Spatial Grasp Technology. Springer, 2021. 305 p.

31. P.S. Sapaty, Complexity in International Security: A Holistic Spatial Approach. Emerald Publishing, 2019. 160 p.
32. P.S. Sapaty, Holistic Analysis and Management of Distributed Social Systems. Springer, 2018. 234 p.
33. P.S. Sapaty, Managing Distributed Dynamic Systems with Spatial Grasp Technology. Springer, 2017. 284 p.
34. P.S. Sapaty, Ruling Distributed Dynamic Worlds. New York: John Wiley & Sons, 2005. 255 p.
35. P.S. Sapaty, Mobile Processing in Distributed and Open Environments. New York: John Wiley & Sons, 1999. 410 p.
36. P.S. Sapaty, Global Network Management under Spatial Grasp Paradigm. *International Robotics & Automation Journal*, 6(3) (2020), 134–148. https://medcraveonline.com/IRATJ/IRATJ-06-00212.pdf
37. P.S. Sapaty, Global Network Management under Spatial Grasp Paradigm, *Global Journal of Researches in Engineering: J General Engineering*, 20(5) (2020), 58–81. https://globaljournals.org/GJRE_Volume20/6-Global-Network-Management.pdf
38. P.S. Sapaty, Advanced terrestrial and celestial missions under spatial grasp technology, *Aeronautics and Aerospace Open Access Journal*, 4(3) (2020). https://medcraveonline.com/AAOAJ/AAOAJ-04-00110.pdf
39. P.S. Sapaty, Spatial Management of Distributed Social Systems, *Journal of Computer Science Research*, 2(3) (2020, July). https://ojs.bilpublishing.com/index.php/jcsr/article/view/2077/pdf
40. P.S. Sapaty, Towards Global Nanosystems under High-level Networking Technology, *Acta Scientific Computer Sciences*, 2(8) (2020). https://www.acta-scientific.com/ASCS/pdf/ASCS-02-0051.pdf
41. P.S. Sapaty, Symbiosis of Distributed Simulation and Control under Spatial Grasp Technology, *SSRG International Journal of Mobile Computing and Application (IJMCA)*, 7(2) (2020, May–August). http://www.internationaljournalssrg.org/IJMCA/2020/Volume7-Issue2/IJMCA-V7I2P101.pdf
42. P.S. Sapaty, Global Network Management under Spatial Grasp Paradigm, *International Robotics & Automation Journal*, 6(3) (2020). https://medcraveonline.com/IRATJ/IRATJ-06-00212.pdf
43. P.S. Sapaty, Global Network Management under Spatial Grasp Paradigm, *Global Journal of Researches in Engineering: J General Engineering*, 20(5) (2020). https://globaljournals.org/GJRE_Volume20/6-Global-Network-Management.pdf
44. P.S. Sapaty, Symbiosis of Virtual and Physical Worlds under Spatial Grasp Technology, *Journal of Computer Science & Systems Biology*, 13(6) (2020). https://www.hilarispublisher.com/open-access/symbiosis-of-virtual-and-physical-worlds-under-spatial-grasp-technology.pdf
45. P.S. Sapaty, Symbiosis of Real and Simulated Worlds under Global Awareness and Consciousness, *The Science of Consciousness* | *TSC* 2020. https://eagle.sbs.arizona.edu/sc/report_poster_detail.php?abs=3696
46. P.S. Sapaty, Spatial Grasp as a Model for Space-based Control and Management Systems, *Mathematical Machines and Systems*, 1 (2021), 135–138. http://www.immsp.kiev.ua/publications/articles/2021/2021_1/Sapaty_book_1_2021.pdf
47. P.S. Sapaty, Managing multiple satellite architectures by spatial grasp technology, *Mathematical Machines and Systems* 1 (2021), 3–16. http://www.immsp.kiev.ua/publications/eng/2021_1/

48. P.S. Sapaty, Spatial Management of Large Constellations of Small Satellites, *Mathematical Machines and Systems* 2 (2021). http://www.immsp.kiev.ua/publications/articles/2021/2021_2/02_21_Sapaty.pdf
49. P.S. Sapaty, Global Management of Space Debris Removal under Spatial Grasp Technology, *Acta Scientific Computer Sciences*, 3(7) (2021, July). https://www.actascientific.com/ASCS/pdf/ASCS-03-0135.pdf
50. P.S. Sapaty, Space Debris Removal under Spatial Grasp Technology, *Network and Communication Technologies*, 6(1) (2021). https://www.ccsenet.org/journal/index.php/nct/article/view/0/45486
51. P.S. Sapaty, Spatial Grasp Model for Management of Dynamic Distributed Systems, *Acta Scientific Computer Sciences*, 3(9) (2021). https://www.actascientific.com/ASCS/pdf/ASCS-03-0170.pdf
52. P.S. Sapaty, Spatial Grasp Model for Dynamic Distributed Systems, *Mathematical Machines and Systems*, 3 (2021). http://www.immsp.kiev.ua/publications/articles/2021/2021_3/03_21_Sapaty.pdf
53. P.S. Sapaty, Development of Space-based Distributed Systems under Spatial Grasp Technology, *Mathematical Machines and Systems*, 4 (2021).

Chapter 9

Space debris removal under Spatial Grasp Technology (SGT)

9.1 INTRODUCTION

There are millions of pieces of space junk flying around the Earth, and especially in low Earth orbits (LEO); see the space junk review in Chapter 2, also [1–16]. The quantity of space junk may be rapidly increasing due to the intensive launch of multi-satellite constellations (sometimes even called "mega-constellations" for their sizes) for very different purposes and by different countries, particularly in LEO, especially when these satellites come to the end of service or collide with other satellites or the existing junk.

In addition to the Chapter 2 debris summary, some theoretical background in this area should be mentioned too. To reduce the rising influence of the space debris and improve the safe performance of the space mission, the three-stage removal strategy for space debris is proposed [17], which can also effectively reduce the fuel consumption. In [18], space debris removal from a game-theoretic perspective is studied, focused on the question how self-interested agents can cooperate. Centralized and decentralized solutions are compared, and the main finding is that the cost of a decentralized, competitive solution could be significant, and should be taken into consideration when forming debris-removal strategies.

In this chapter, we are considering different possibilities of organizing massive removal of most dangerous junk items, especially in the LEO orbits, by large constellations of cleaning satellites, which may be the only viable solution. The cleaning constellations are supposed to be operating under the developed high-level Spatial Grasp Technology (SGT), described in Chapters 3–4 and many existing publications [19–35].

The rest of this chapter is organized as follows. *Section 9.2* provides a brief summary of SGT described in detail in Chapters 3 and 4. *Section 9.3* shows how a special constellation of junk cleaning satellites can be converted under SGT into an intelligent team capable of organizing and

DOI: 10.1201/9781003230090-9

executing massive debris removal operations with autonomy and reduced ground communications. *Section 9.4* shows how the debris discovery, general organization, and execution scenarios can be expressed in Spatial Grasp Language (SGL), which is interpreted in a distributed way by the whole satellite cleaning network, along with showing how the removal solutions can also be organized simultaneously for multiple debris. *Section 9.5* provides a solution in SGL where the junk items can be virtually treated as active objects traveling around the Earth and finding proper cleaners by their own initiative, and this allows any time needed for finding a suitable match for such self-removal. *Section 9.6* shows how this strategy can be effectively integrated with the one in the previous section. *Section 9.7* concludes the chapter.

9.2 A BRIEF SUMMARY OF SPATIAL GRASP TECHNOLOGY

Within Spatial Grasp Technology (SGT), described in Chapters 3–4, also in more detail in [19–25], a high-level scenario for any task to be performed in a distributed world is represented as an *active self-evolving pattern*, rather than a traditional program, sequential or parallel one. This pattern, written in a recursive high-level SGL and expressing top semantics of the problem to be solved can start from any point of the world. Then it spatially *propagates, replicates, modifies, covers, and matches* the distributed world in a parallel wavelike mode, while echoing the reached control states and data found or obtained for making decisions at higher levels and further space navigation, as symbolically shown in Figure 9.1.

Many spatial processes in SGL can start any time and in any place, cooperating or competing with each other, depending on applications. The self-spreading and self-matching SGL patterns-scenarios can create *active spatial infrastructures* covering any regions. These infrastructures can effectively support or express distributed knowledge bases, advanced command and control, situation awareness, autonomous and collective decisions, as well as any existing or hypothetical computational and/or control models, systems, and solutions. SGL has a deep recursive structure with its parallel

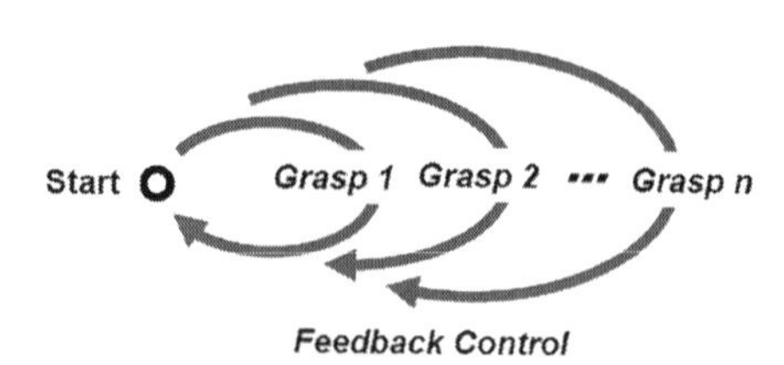

Figure 9.1 Controlled wavelike coverage and conquest of distributed spaces by SGT.

spatial scenarios called *grasps* and universal operational and control constructs as *rules* (braces identifying repetition):

```
grasp → constant | variable | rule ({ grasp,})
```

9.3 INTELLIGENT CONSTELLATION OF JUNK CLEANERS

In Chapters 5–8, also [27–35], organization of constellations of small satellites was considered under SGT with their applications for very different purposes, where large satellite groups can become intelligent, self-organized systems capable of solving very complex problems autonomously, also with reduced engagement of costly ground antennas and their infrastructures. Dealing with such complex problems as huge amount of space debris can also be possible only by using large constellations, even mega-ones, of special cleaning (like de-orbiting) satellites working together as a global goal-oriented system. In Figure 9.2, such constellation of cleaners is symbolically shown with all units supplied with mechanical grasps, although they may also be equipped with any other techniques (including the ones mentioned in Chapter 2).

In the following sections we will be showing and discussing some elementary scenarios in SGL for cooperating cleaning procedures executed by such satellite groups.

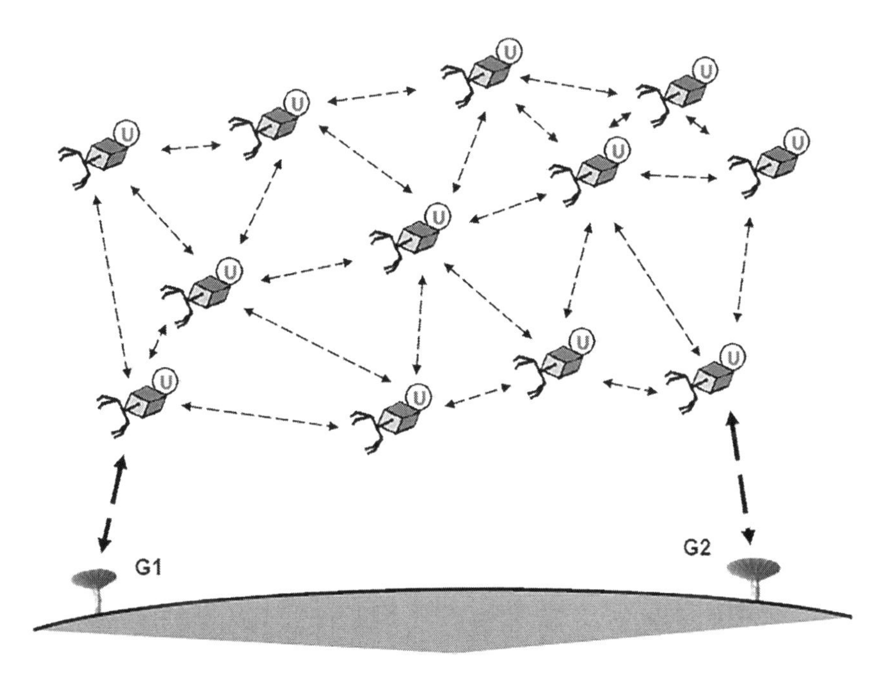

Figure 9.2 Converting constellation of junk cleaners into a self-organized system.

9.4 JUNK REMOVAL BY THE INITIATIVE OF CLEANERS NETWORK

9.4.1 Finding and de-orbiting a single junk item

It is supposed that ground station G2 using radar discovers a suitable junk item and records its parameters (incl. current time, location, size, expected orbit, etc.) in frontal variable `Details`, and then transfers this to the nearest satellite—cleaner `C1`, by the radar too, as in the following SGL scenario. (See also Figure 9.3, with the related debris item named as `D1`.) The chosen satellite starts repetitive and parallel constellation flooding and coverage (up to the whole network) using direct links between satellites until it finds the most suitable cleaner for the junk de-orbiting (i.e., its `Snapshot` of the nearest satellite matches `Details`). If de-orbiting is provided successfully, further constellation coverage is aborted with appropriate cleaning of the network (operation `abort`). Otherwise, the network will continue searching for the appropriate cleaner. As the network search takes time, with satellites constantly moving, and the junk items moving too, the networked coverage scenario regularly updates the expected for this moment of time (with use of global variable `TIME`) junk parameters in the moving variable `Details`. Operation `hop_first` navigates the cleaners' network

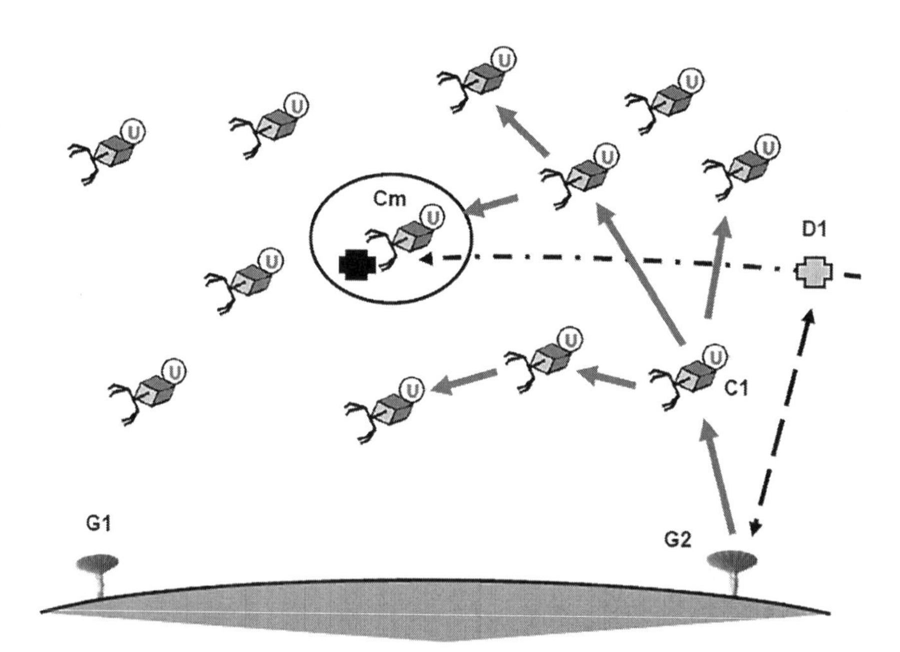

Figure 9.3 Autonomous reaching of the junk needed by self-organized network of cleaners.

with blocking possible cycling (using nodes marking on the internal interpretation level).

```
hop(G2);
frontal(Details) = find_select(radar, junk, TIME);
hop_first(any_cleaner, radar);
repeat(
   Snapshot = parameters(closest_junk, seen);
   if(match(Details, Snapshot),
      (deorbit(Snapshot); abort));
   update(Details, TIME);
   hop_first(all_cleaners, direct_links))
```

9.4.2 Removing many junk items simultaneously

Any number of registered junk items can be found by the ground stations with launching de-orbiting processes simultaneously, as by the following scenario (see also Figure 9.4) using two ground stations `G1` and `G2`, each trying to find and remove suitable junks independently by the same network of cleaners. This navigation independence within the same satellite network is guaranteed by different navigation colors using frontal environmental variables `IDENTITY` (here corresponding to names `G1` and `G2`), which also influence the names of used SGL variables and internal blocking of cycling. In a further development, each ground station can also detect and record any number of suitable junk items for the removal attempts (and not only

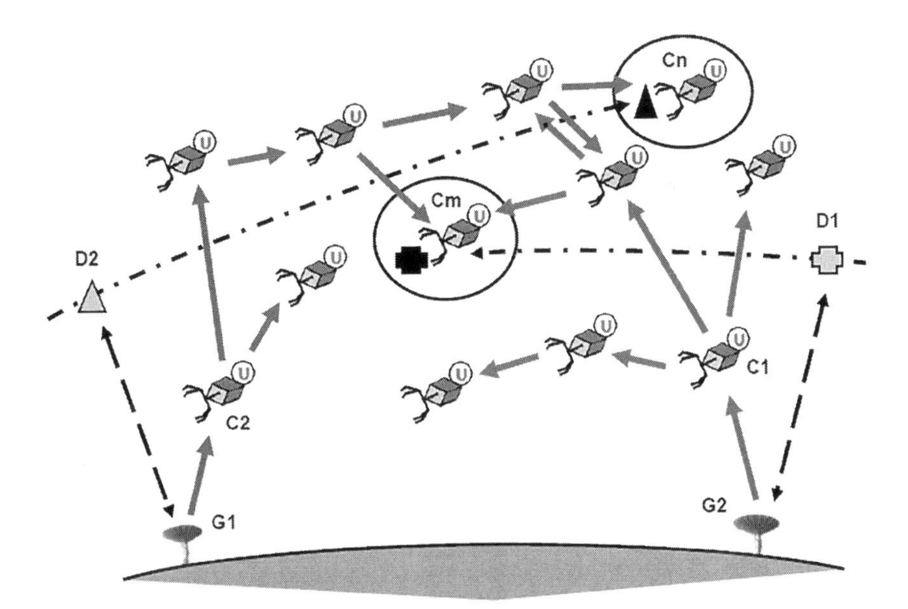

Figure 9.4 Simultaneous junk reaching and removal.

a single one for each station, as shown in Figure 9.4 as D1 and D2), which can trigger independent network navigation with individual colors for each junk item too.

```
hop(G1, G2); IDENTITY = NAME;
frontal(Details) = find_select(radar, junk, TIME);
hop_first(any_cleaner, radar);
repeat(
   Snapshot = parameters(closest_junk, seen);
   if(match(Details, Snapshot),
      (deorbit(Snapshot); abort));
   update(Details, TIME);
   hop_first(all_cleaners, direct_links))
```

9.5 ACTIVE VIRTUAL JUNK SOLUTION

In the previous section we used parallel network search for a possible match between the selected junk and the cleaning satellite which happened to be close enough to perform de-orbiting operation. If the latter is unsuccessful, the parallel network flooding would be continued until full network coverage, which may not always guarantee the final success as satellites and junk may happen to be far away from each other during this period of time, on different orbits, and moving in different directions. The following scenario (see also Figure 9.5) is putting the global search for the match not on the initiative of the cleaners network but on the activity of the junk item itself by having represented it as a virtual identity process traveling between distributed cleaners any needed time, including endlessly, with possibility of entering same cleaners many times unless one of them happens to match the needed junk item at some moment of time. And such virtual junk can be traveling only via local neighbors, which appears more suitable for the continuing search without expensive parallel flooding of the whole network. It is supposed from the beginning that the Details of the junk item of interest are obtained by some ground station, which sends them directly to the discovered nearest cleaner C1, from which the virtual junk identity process begins traveling autonomously through the cleaners' constellation, as long as needed, and around the Earth.

```
frontal(Details = junk_data, Snapshot);
hop(C1);
repeat(
   repeat(
     Snapshot = parameters(closest_junk, seen);
     match(Details, Snapshot);
     if(deorbit(Snapshot), abort);
     update(Details, TIME));
   or(maxdestination(
        hop_neighbors(all_cleaners, direct_links);
        closeness(Details, junk_seen) > threshold),
     stay))
```

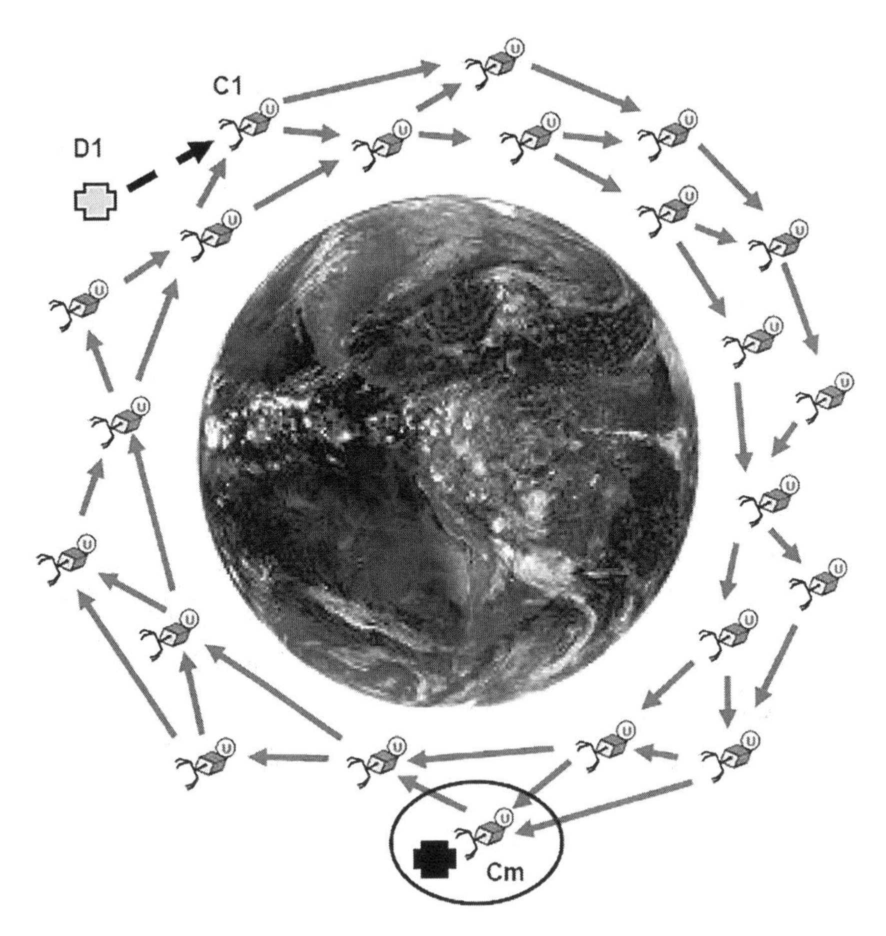

Figure 9.5 Active virtual junk item self-searching for its removal.

9.6 COMBINED PHYSICAL-VIRTUAL ORGANIZATION

It may be useful to combine the two different strategies discussed above where first one was the matching search on the initiative and activity of the whole cleaners' network by its global self-flooding, and the other was unlimited localized search for the appropriate cleaner by the initiative of virtual junk itself. The following scenario combines the global search for initial approximate matching (similar to Figures 9.3 and 9.4), starting from some ground station G2, after which the junk is delegated its own initiative to find absolute match with any time needed for this (as in Figure 9.5).

```
hop(G2);
frontal(Details) = find_select(radar, junk, TIME);
hop_first(any_cleaner, radar);
```

```
repeat(
   Snapshot = parameters(closest_junk, seen);
   if(close(Details, Snapshot),
      repeat(
         repeat(
           Snapshot = parameters(closest_junk, seen);
           match(Details, Snapshot);
           if(deorbit(Snapshot), abort);
           update(Details, TIME));
         or(maxdestination(
              hop_neighbors(all_cleaners, direct_links);
              closeness(Details, junk_seen) > threshold))
            stay)));
   update(Details, TIME);
   hop_first(all_cleaners, direct_links))
```

In a further development, we can use such symbiosis for initial discovery of any junk items, their approximate matching with cleaners, and then converting into active virtual junk items, themselves traveling and finding the final solutions, as long and far away as this may be needed.

9.7 CONCLUSIONS

We have considered different possibilities of organization of massive removal of most dangerous junk items, especially in the LEO orbits, by special large constellations of cleaning satellites, which can be effectively organized as intelligent autonomous systems under the developed Spatial Grasp Model and Technology. Of course, such mega-cleaning constellations do not exist yet, and unfortunately many private companies and governmental organizations are chaotically launching numerous and cheap satellites (which soon will count up to 100,000 in LEO). But the danger of uncontrolled rubbishing of space around Earth may be very high and can lead to even more severe consequences than the global warming and current worldwide COVID disaster. So we hope that this book, and the current chapter particularly, will be stimulating and useful ahead of creating global cleaning approaches, and we also plan to offer and discuss these issues with reputable international organizations.

REFERENCES

1. Space Debris, NASA Headquarters Library. https://www.nasa.gov/centers/hq/library/find/bibliographies/space_debris
2. A.C. Boley, M. Byers, Satellite Mega-Constellations Create Risks in Low Earth Orbit, The Atmosphere and on Earth, *Scientific Report*. https://www.nature.com/articles/s41598-021-89909-7

3. Research on space debris, safety of space objects with nuclear power sources on board and problems relating to their collision with space debris. Committee on the Peaceful Uses of Outer Space. Vienna, 2019. http://www.unoosa.org/res/oosadoc/data/documents/2019/aac_105c_12019crp/aac_105c_12019crp_7_0_html/AC105_C1_2019_CRP07E.pdf
4. Deorbit Systems, National Aeronautics and Space Administration, November 28, 2020. https://www.nasa.gov/smallsat-institute/sst-soa-2020/passive-deorbit-systems
5. A. Froehlich (Ed.), *Space Security and Legal Aspects of Active Debris Removal*, Springer, 2019. https://www.springer.com/gp/book/9783319903378
6. A. Sheer, S. Li, Space Debris Mounting Global Menace Legal Issues Pertaining to Space Debris Removal: Ought to Revamp Existing Space Law Regime, *Beijing Law Review*, 10 (2019), 423–440. https://www.scirp.org/pdf/BLR_2019051615104007.pdf
7. G.S. Aglietti, From Space Debris to NEO, Some of the Major Challenges for the Space Sector, *Frontiers of Space Technology*, 16 (2020, June). https://www.frontiersin.org/articles/10.3389/frspt.2020.00002/full
8. L. David, Space Junk Removal Is Not Going Smoothly, *Scientific American* 14 (2021, April). https://www.scientificamerican.com/article/space-junk-removal-is-not-going-smoothly/
9. ESA Commissions World's First Space Debris Removal, *ESA / Safety & Security / Clean Space*, 9 December, 2019. https://www.esa.int/Safety_Security/Clean_Space/ESA_commissions_world_s_first_space_debris_removal
10. A. Parsonson, ESA Signs Contract for First Space Debris Removal Mission, *Space News*, 2 December, 2020. https://spacenews.com/clearspace-contract-signed/
11. Humza, The First-Ever Space Mission to Clean Orbital Junk Will Use a Giant Claw, *Techspot*, 1 December, 2020. https://www.techspot.com/community/topics/the-first-ever-space-mission-to-clean-orbital-junk-will-use-a-giant-claw.266509/
12. Japanese Company Planning Space Debris Removal by Laser on Satellite, *Kyodo News*, 8 August, 2020. https://english.kyodonews.net/news/2020/08/fc06829d1d9a-japanese-company-planning-space-debris-removal-by-laser-on-satellite.html
13. Plasma Thruster: New Space Debris Removal Technology, Tohoku University, 27 September, 2018. https://www.eurekalert.org/pub_releases/2018-09/tu-ptn092718.php
14. K. Hunt, Mission to Clean Up Space Junk with Magnets Set for Launch, CNN, 1 April, 2021. https://edition.cnn.com/2021/03/19/business/space-junk-mission-astroscale-scn/index.html
15. M. Obe, Japan's Astroscale Launches Space Debris-Removal Satellite, Nikkei Asia, 22 March, 2021. https://asia.nikkei.com/Business/Aerospace-Defense/Japan-s-Astroscale-launches-space-debris-removal-satellite
16. C. Weiner, New Effort to Clean up Space Junk Reaches Orbit, 21 March, 2021. https://www.npr.org/2021/03/21/979815691/new-effort-to-clean-up-space-junk-prepares-to-launch
17. Y. Chen et al., Optimal mission planning of active space debris removal based on genetic algorithm, *IOP Conference Series: Materials Science and Engineering* 715 (2020), 012025. https://iopscience.iop.org/article/10.1088/1757-899X/715/1/012025/pdf

18. R. Klima et al., Space Debris Removal: Learning to Cooperate and the Price of Anarchy, *Frontiers in Robotics and AI*, 4 June, 2018. https://www.frontiersin.org/articles/10.3389/frobt.2018.00054/full
19. P.S. Sapaty, Symbiosis of Real and Simulated Worlds under Spatial Grasp Technology, Springer, 2021. 251 p.
20. P.S. Sapaty, Complexity in International Security: A Holistic Spatial Approach, Emerald Publishing, 2019. 160 p.
21. P.S. Sapaty, Holistic Analysis and Management of Distributed Social Systems, Springer, 2018. 234 p.
22. P.S. Sapaty, Managing Distributed Dynamic Systems with Spatial Grasp Technology, Springer, 2017. 284 p.
23. P.S. Sapaty, Ruling Distributed Dynamic Worlds. John Wiley & Sons, 2005. 255 p.
24. P.S. Sapaty, Mobile Processing in Distributed and Open Environments. John Wiley & Sons, 1999. 410 p.
25. P.S. Sapaty, A distributed processing system, European Patent N 0389655, Publ. 10.11.93, European Patent Office. 35 p.
26. P.S. Sapaty, Global Network Management under Spatial Grasp Paradigm, *Global Journal of Researches in Engineering: J General Engineering* 20(5) (2020), 58–81. https://globaljournals.org/GJRE_Volume20/6-Global-Network-Management.pdf
27. P.S. Sapaty, Advanced Terrestrial and Celestial Missions under Spatial Grasp Technology. *Aeronautics and Aerospace Open Access Journal* 4(3) (2020). https://medcraveonline.com/AAOAJ/AAOAJ-04-00110.pdf.
28. P.S. Sapaty, Managing Multiple Satellite Architectures by Spatial Grasp Technology, *Mathematical Machines and Systems* 1 (2021), 3–16. http://www.immsp.kiev.ua/publications/eng/2021_1/
29. P.S. Sapaty, Spatial Management of Large Constellations of Small Satellites, *Mathematical Machines and Systems* 2 (2021). http://www.immsp.kiev.ua/publications/articles/2021/2021_2/02_21_Sapaty.pdf
30. P.S. Sapaty, Global Management of Space Debris Removal under Spatial Grasp Technology, *Acta Scientific Computer Sciences* 3(7) (2021, July). https://www.actascientific.com/ASCS/pdf/ASCS-03-0135.pdf
31. P.S. Sapaty, Global Network Management under Spatial Grasp Paradigm, *International Robotics & Automation Journal* 6(3) (2020), 134–148. https://medcraveonline.com/IRATJ/IRATJ-06-00212.pdf
32. P.S. Sapaty, Space Debris Removal under Spatial Grasp Technology, *Network and Communication Technologies* 6(1) (2021). https://www.ccsenet.org/journal/index.php/nct/article/view/0/45486
33. P.S. Sapaty, Spatial Grasp Model for Management of Dynamic Distributed Systems, *Acta Scientific Computer Sciences* 3(9) (2021). https://www.actascientific.com/ASCS/pdf/ASCS-03-0170.pdf
34. P.S. Sapaty, Spatial Grasp Model for Dynamic Distributed Systems, *Mathematical Machines and Systems* 3 (2021). http://www.immsp.kiev.ua/publications/articles/2021/2021_3/03_21_Sapaty.pdf
35. P.S. Sapaty, Development of Space-based Distributed Systems under Spatial Grasp Technology, *Mathematical Machines and Systems* 4 (2021).

Chapter 10

Conclusions

10.1 INTRODUCTION

This chapter concludes the book, which has offered a unified approach toward organization of large satellite constellations, especially on Low Earth Orbits [1–42], and at any stages of their development, starting from accidental launches to gradual growth to maturity. Such constellations at the beginning may not have stable communication structures and network topologies similar to the existing terrestrial systems. The management approach offered with wavelike mobile spatial code covering distributed systems in a spatial pattern-matching mode is ideologically stemming from the organization and implementation of first citywide heterogeneous computer networks half a century ago [43–52], well before the internet. It has grown up into a high level networking technology [53–123] that can effectively cover the rapidly growing celestial systems, from elementary communication protocols to high-level holistic solutions of complex problems, both civil and military. Traditional parallel and distributed system representations as parts or agents operating on individual algorithms and exchanging messages with each other are becoming inadequate, as such systems can be difficult to outline ahead of the solutions needed from them, and this especially relates to dynamic and rapidly growing systems around the Earth. These systems may need much stronger and quicker integration, often particularly task-oriented and on the fly, for solving emergent and nonpredictable-in-advance problems, and this is the main orientation and aim of this book.

10.2 MAIN BOOK RESULTS

The main findings and achievements of this book can be summarized as follows.

1. A ***substantial review has been made on the past and ongoing space activity***, which covered the growing satellite constellations; past, current, and planned defense and industrial projects; and rapidly growing

DOI: 10.1201/9781003230090-10

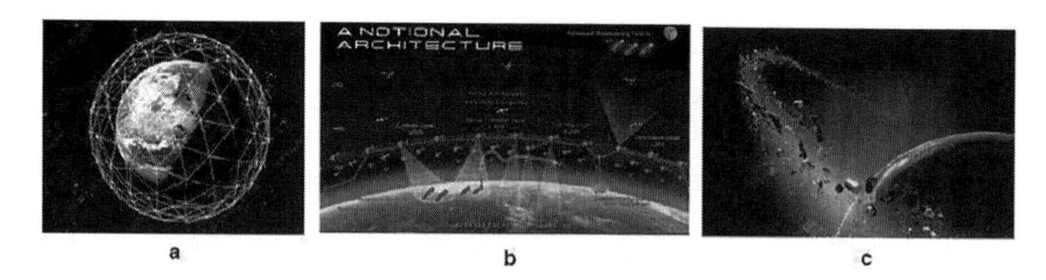

Figure 10.1 Areas of investigation: (a) satellite constellations; (b) new space architectures; (c) space debris.

space debris which can endanger any further use of space. This was in Chapter 2, see also existing publications on these topics [1–42] and Figure 10.1 summarizing the main areas of book's investigation and research.

The book main practical results and achievements are mentioned further down, see also Figure 10.2 summarizing and showing together related key figures from different chapters.

2. A ***new computational, control, and management Spatial Grasp model has been offered*** as a natural extension of traditional concept of algorithm with potential applications in large distributed systems operating in combined terrestrial and celestial environments. A resultant updated version of Spatial Grasp Technology (SGT) has been described within which a high-level scenario in a Spatial Grasp Language (SGL) is represented as an active self-evolving pattern rather than a traditional program. This pattern, starting from any point, spatially propagates, replicates, modifies, covers and matches the distributed world in parallel wavelike mode, while echoing the reached control states and data found or obtained for making decisions at higher levels and further space navigation. Many spatial processes in SGL can start any time and in any place, cooperating or competing with each other, depending on applications. This was described in Chapters 3 and 4, with more details in [65–107], see also Figure 10.2(a).
3. It was ***shown how to convert large constellations into capable self-organized systems*** under SGT by supplying satellites with SGL interpreters which can communicate directly and also with ground stations, and this may also result in significant simplification of ground antennas and reduction of their numbers. Some basic operations over satellite constellations in a virus-like self-spreading parallel mode were demonstrated in SGL, which included broadcasting executive orders to all satellites via changeable communications between them, collecting and returning data accumulated by all satellites, and constellation repositioning and restructuring. This was in Chapter 5, with the related publications in [116–118], see also Figure 10.2(b).

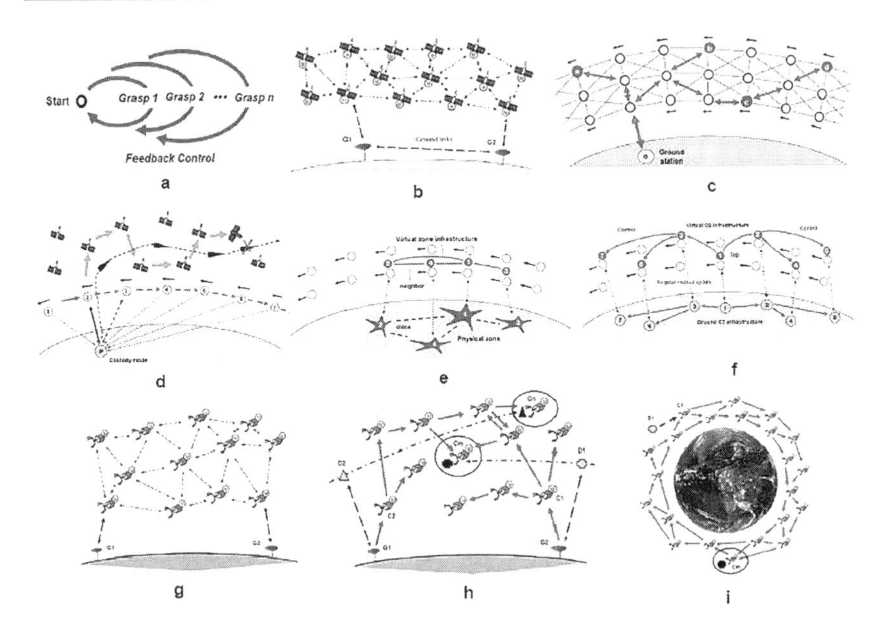

Figure 10.2 Main achievements of the book: (a) New version of Spatial Grasp Technology (SGT); (b) converting satellite constellations into self-organized system by SGT; (c) managing transport layer under SGT; (d) integrating tracking and custody layers by SGT; (e) introducing virtual layer overseeing custody infrastructures by SGT; (f) using virtual layer for distributed command and control by SGT; (g) converting cleaning constellation into autonomous system by SGT; (h) simultaneous debris removal in SGT; (i) converting junk items into self-removing virtual objects by SGT.

4. ***The transport layer***, which is basic for any activities and projects using satellite constellations, being also a key one in SDA architecture [14–20], ***was analyzed for its organization and management*** by SGT. It was shown how to deal with highly dynamic constellation topologies for reaching proper satellite nodes, delivery of a package to proper nodes and collection and return certain items from them, also how to introduce special measures for dealing with high constellation unpredictability and dynamics. It was demonstrated how to create and fix communication topology in case of stable distances between satellites, and how to express in SGL such basic networking operations as finding shortest paths and routing tables, also discovering and analyzing certain components and structures in distributed networks with stable topologies, like articulation points and cliques. This was in Chapter 6, with related publications in [14–26], see also Figure 10.2(c).
5. The main problem with cheap and relatively simple LEO satellites is that they are rapidly changing their positions over Earth

locations, and to provide continuous observation, they regularly need to transfer their duties and accumulated information to other satellites. It was ***shown how this can be effectively managed by the super-virus-like SGT and its basic SGL language***, which navigate multi-satellite constellations in parallel virus-like mode, by describing exemplary solutions for multi-satellite projects, like SDI brilliant pebbles from the past and also new SDA space architecture. For the latter, it demonstrated how self-spreading SGL scenarios can effectively track and destroy complexly moving objects like hypersonic gliders, how to implement custody layer for constant monitoring of certain terrestrial objects, and how to organize cooperative work of custody and tracking layers by SGT under stable constellation topologies. This was in Chapter 7, see also [14–23], and Figure 10.2(d).

6. It was shown ***how to introduce a special virtual layer for satellite constellations*** which may be over and superior to any other layers. By introducing such layer, it became possible to organize advanced custody not only for particular objects on Earth but also for any large distributed terrestrial infrastructures. It was demonstrated how such layer helped in analyzing terrestrial systems like large zones with multiple fires, and worldwide delivery of goods. Such layer also helps in implementing distributed virtual-physical command and control with regularly updated correspondence between virtual and physical (or earth-based) C2. It was shown how to manage two- and then arbitrary multi-level virtual-physical C2 infrastructure, which can, if needed, shift global control of large operations and missions partially or completely into space, say, in case of critical situations on Earth. This was in Chapter 8, see also [14–23], and Figures 10.2(e, f).
7. It was shown how a special ***constellation of junk cleaning satellites can be converted under SGT into an intelligent team*** capable of organizing and executing massive debris-removal operations with autonomy and reduced ground communications. This discovery, general organization, and execution scenarios can be totally expressed in SGL, which is interpreted in a distributed way by the whole satellite cleaning network, where it is also shown how the removal can also be organized simultaneously for multiple debris. A solution was demonstrated in SGL where the junk items can be virtually treated as active objects traveling around the Earth and finding proper cleaners by their own initiative, and this may allow any time needed, actually unlimited, for finding a suitable match for their self-removal. It was also shown how different junk-removal strategies can effectively be combined by SGT. This was in Chapter 9, see also [27–42], and Figure 10.2(g–i).

10.3 OTHER POSSIBILITIES PROVIDED BY SPATIAL GRASP TECHNOLOGY

The below-mentioned other areas and topics were investigated and tested under SGT and its previous versions, with many details available in [53–123].

a) Distributed Interactive simulation of defense-oriented systems
b) Social systems
c) Security and defense
d) Electronic warfare
e) Mosaic Warfare
f) Collective robotics
g) Gestalt philosophy, theory, and holistic systems
h) Artificial Intelligence
i) Consciousness and global awareness
j) Classical graph and network theory
k) New high-level networking protocols
l) Simulation of world pandemics and its fighting methods
m) Spatial art

More research and development continues in these and other fields under the Spatial Grasp philosophy, model, and technology, with new patents, publications, books including, being planned and expected.

10.4 TECHNOLOGY IMPLEMENTATION ISSUES

The described latest SGT version, suitable for the advanced constellations management, can be quickly implemented on any platform, even within traditional university environments, similar to the previous technology versions implemented and tested in different countries under the author's supervision. The author will also be happy to provide any support needed for the new implementations, including cooperative preparation and writing of different application projects in SGL or its modified versions which may be closer to the particular projects.

REFERENCES

1. N. Mohanta, How many satellites are orbiting the Earth in 2021? *Geospatial World* 28 May, 2021. https://www.geospatialworld.net/blogs/how-many-satellites-are-orbiting-the-earth-in-2021/
2. Space Debris, NASA Headquarters Library, https://www.nasa.gov/centers/hq/library/find/bibliographies/space_debris

3. United Nations Register of Objects Launched into Outer Space, The United Nations Office for Outer Space Affairs. http://www.unoosa.org/oosa/en/spaceobjectregister/index.html
4. G. Martin, NewSpace: The «Emerging» Commercial Space Industry. https://ntrs.nasa.gov/archive/nasa/casi.ntrs.nasa.gov/20140011156.pdf
5. J.-M. Bockel, The Future of the Space Industry. General Report. 17 November, 2018. https://www.nato-pa.int/download-file?filename=sites/default/files/2018-12/2018%20-%20THE%20FUTURE%20OF%20SPACE%20INDUSTRY%20-%20BOCKEL%20REPORT%20-%20173%20ESC%2018%20E%20fin.pdf
6. G. Curzi, D. Modenini, P. Tortora, Review Large Constellations of Small Satellites: A Survey of Near Future Challenges and Missions, *Aerospace* 7(9) (2020), 133. https://www.mdpi.com/2226-4310/7/9/133/htm
7. A. Venkatesan, J. Lowenthal, P. Prem, M. Vidaurri, The impact of satellite constellations on space as an ancestral global commons, *Nature Astronomy* 4 (2020, November), 1043–1048. www.nature.com/natureastronomy
8. R. Skibba, How satellite mega-constellations will change the way we use space, *MIT Technology Review* 26 February, 2020. https://www.technologyreview.com/2020/02/26/905733/satellite-mega-constellations-change-the-way-we-use-space-moon-mars/
9. M. Minet, The Space Legal Issues with Mega-Constellations, 3 November, 2020. https://www.spacelegalissues.com/mega-constellations-a-gordian-knot/
10. E. Siegel, Astronomy Faces a Mega-Crisis as Satellite Mega-Constellations Loom, 19 January, 2021. https://www.forbes.com/sites/startswithabang/2021/01/19/astronomy-faces-a-mega-crisis-as-satellite-mega-constellations-loom/?sh=30597dca300d
11. A. Jones, China is developing plans for a 13,000-satellite megaconstellation, *Space News*, 21 April, 2021. https://spacenews.com/china-is-developing-plans-for-a-13000-satellite-communications-megaconstellation/
12. N. Reilanda, A. J. Rosengren, R. Malhotra, C. Bombardelli, Assessing and minimizing collisions in satellite mega-constellations, 2021, Published by Elsevier B.V. on behalf of COSPAR. https://www.sciencedirect.com/science/article/abs/pii/S0273117721000326
13. J. Fous, Mega-constellations and mega-debris, October 10, 2016. https://www.thespacereview.com/article/3078/1
14. Space Development Agency Next-Generation Space Architecture. 2019. https://www.airforcemag.com/PDF/DocumentFile/Documents/2019/SDA_Next_Generation_Space_Architecture_RFI%20(1).pdf
15. S. Magnuson, Web Exclusive: Details of the Pentagon's New Space Architecture Revealed. 2019. https://www.nationaldefensemagazine.org/articles/2019/9/19/details-of-the-pentagon-new-space-architecture-revealed
16. D. Messier, Space Development Agency Seeks Next-Gen Architecture in First RFI. 2019. http://www.parabolicarc.com/2019/07/07/space-development-agency-issues-rfi/
17. V. Insinna, Space agency has an ambitious plan to launch 'hundreds' of small satellites. Can it get off the ground? *DefenceNews*. Space. 10 April, 2019. https://www.defensenews.com/space/2019/04/10/sda-has-an-ambitious-plan-to-launch-hundreds-of-small-satellites-can-it-get-off-the-ground/

18. N. Strout, Space Development Agency approves design for satellites that can track hypersonic weapons, C4ISRNET, 2021. https://www.c4isrnet.com/battlefield-tech/space/2021/09/20/space-development-agency-approves-design-for-satellites-that-can-track-hypersonic-weapons/
19. Transport Layer Tranche-0 A-Class 1, ..., 7. https://space.skyrocket.de/doc_sdat/transport-layer-tranche-0-a-class-york.htm
20. Transport, SDA. https://www.sda.mil/transport/
21. Custody, SDA, https://www.sda.mil/custody/
22. Battle Management Command, Control, and Communication (BMC3), SDA. https://www.sda.mil/battle-management/
23. Transport Layer. https://en.wikipedia.org/wiki/Transport_layer
24. M.A.A. Madni, S. Iranmanesh, R. Raad, Review DTN and Non-DTN Routing Protocols for Inter-CubeSat Communications: A comprehensive survey. *Electronics* 9 (2020), 482. https://www.mdpi.com/2079-9292/9/3/482
25. Q.I. Xiaogang, M.A. Jiulong, W.U. Dan, L.I.U. Lifang, H.U. Shaolin. A survey of routing techniques for satellite networks. *Journal of Communications and Information Networks* 1(4) (2016), 66–85. http://www.infocomm-journal.com/jcin/EN/10.11959/j.issn.2096-1081.2016.058
26. M.A.A. Madni, S. Iranmanesh, R. Raad, Review DTN and non-DTN routing protocols for inter-CubeSat communications: A comprehensive survey. *Electronics* 9 (2020), 482. https://www.mdpi.com/2079-9292/9/3/482
27. Research on space debris, safety of space objects with nuclear power sources on board and problems relating to their collision with space debris. Committee on the Peaceful Uses of Outer Space. Vienna, 2019. http://www.unoosa.org/res/oosadoc/data/documents/2019/aac_105c_12019crp/aac_105c_12019crp_7_0_html/AC105_C1_2019_CRP07E.pdf
28. Deorbit Systems, National Aeronautics and Space Administration, 28 November, 2020. https://www.nasa.gov/smallsat-institute/sst-soa-2020/passive-deorbit-systems
29. A. Froehlich (Ed.), *Space Security and Legal Aspects of Active Debris Removal*, Springer, 2019. https://www.springer.com/gp/book/9783319903378
30. A. Sheer, S. Li, Space Debris Mounting Global Menace Legal Issues Pertaining to Space Debris Removal: Ought to Revamp Existing Space Law Regime, *Beijing Law Review* 10 (2019), 423–440. https://www.scirp.org/pdf/BLR_2019051615104007.pdf
31. G.S. Aglietti, From Space Debris to NEO, Some of the Major Challenges for the Space Sector, *Frontiers of Space Technology*, 16 June, 2020. https://www.frontiersin.org/articles/10.3389/frspt.2020.00002/full
32. L. David, Space Junk Removal Is Not Going Smoothly, *Scientific American*, 14 April, 2021. https://www.scientificamerican.com/article/space-junk-removal-is-not-going-smoothly/
33. ESA commissions world's first space debris removal, ESA/Safety & Security/Clean Space, 9 December, 2019. https://www.esa.int/Safety_Security/Clean_Space/ESA_commissions_world_s_first_space_debris_removal
34. A. Parsonson, ESA signs contract for first space debris removal mission, *Space News*, 2 December, 2020. https://spacenews.com/clearspace-contract-signed/
35. Humza, The first-ever space mission to clean orbital junk will use a giant claw, *Techspot*, 1 December, 2020. https://www.techspot.com/community/topics/the-first-ever-space-mission-to-clean-orbital-junk-will-use-a-giant-claw.266509/

36. Japanese company planning space debris removal by laser on satellite, *Kyodo News*, 8 August, 2020. https://english.kyodonews.net/news/2020/08/fc06829d1d9a-japanese-company-planning-space-debris-removal-by-laser-on-satellite.html
37. Plasma thruster: New space debris removal technology, *Tohoku University*, 27 September, 2018. https://www.eurekalert.org/pub_releases/2018-09/tu-ptn092718.php
38. K. Hunt, Mission to clean up space junk with magnets set for launch, *CNN*, 1 April, 2021. https://edition.cnn.com/2021/03/19/business/space-junk-mission-astroscale-scn/index.html
39. M. Obe, Japan's Astroscale launches space debris-removal satellite, *Nikkei Asia*, 22 March, 2021. https://asia.nikkei.com/Business/Aerospace-Defense/Japan-s-Astroscale-launches-space-debris-removal-satellite
40. C. Weiner, New Effort To Clean Up Space Junk Reaches Orbit, 21 March, 2021. https://www.npr.org/2021/03/21/979815691/new-effort-to-clean-up-space-junk-prepares-to-launch
41. Y. Chen et al., Optimal mission planning of active space debris removal based on genetic algorithm, *IOP Conf. Series: Materials Science and Engineering* 715 (2020), 012025. https://iopscience.iop.org/article/10.1088/1757-899X/715/1/012025/pdf
42. R. Klima et al., Space Debris Removal: Learning to Cooperate and the Price of Anarchy, *Frontiers in Robotics and AI*, 4 June, 2018. https://www.frontiersin.org/articles/10.3389/frobt.2018.00054/full
43. A.T. Bondarenko, S.B. Mikhalevich, A.I. Nikitin, P.S. Sapaty, Software of BESM-6 computer for communication with peripheral computers via telephone channels, *Computer Software*, Vol. 5, Inst. of Cybernetics Press, Kiev, 1970 (in Russian).
44. A.T. Bondarenko, V.P. Karpus, S.B. Mikhalevich, A.I. Nikitin, P.S. Sapaty, "Information-computing system ABONENT", Tech. Report No. B178338, All-Union Scientific and Technical Inform. Centre, Moscow, 1972 (in Russian).
45. P.S. Sapaty, "A Method of organization of an intercomputer dialogue in the radial computer systems", in *The Design of Software and Hardware for Automatic Control Systems*, Institute of Cybernetics Press, Kiev, 1973 (in Russian).
46. J.G. Grigorjev, V.P. Karpus, L.I. Pristupa, P.S. Sapaty, "Management of a dialogue in the MIR-2--BESM-6 system", in *Proceedings of the Republic Conference Hardware and Software for Management of Dialogue in Computer Systems*, Kiev, 1973 (in Russian).
47. A.T. Bondarenko, S.B. Mikhalevich, P.S. Sapaty, "Intercomputer dialogue in high-level languages", in *Proceedings of the Republic Conference Hardware and Software for Management of Dialogue in Computer Systems*, Kiev, 1973 (in Russian).
48. P.S. Sapaty, "Asynchronous parallel evolvent of nonlinear and control algorithms", in *Software for Control and Information Systems*, Moscow House of Scientific and Technical Publicity, 1973 (in Russian).
49. P.S. Sapaty, "On possibilities of the organization of a direct intercomputer dialogue in ANALYTIC and FORTRAN languages", Publ. No. 74–29, Institute of Cybernetics Press, Kiev, 1974 (in Russian).

50. D.V. Karachenets, E.P. Pozdnjakov, V.N. Moroz, P.S. Sapaty, "Questions of using dialogue in a computer-aided analysis and design of complex technological systems", in *Proceedings of the Republic Conference The Dialogue Tools for Solving Engineering Problems*, Kiev, 1974, (in Russian).
51. P.S. Sapaty, "A method of fast dispatching for parallel execution of tasks", in *System Programming Languages and Methods of Their Implementation*, Institute of Cybernetics Press, Kiev, 1974 (in Russian).
52. P.S. Sapaty, "Solving branching and cycling tasks on multiprocessor systems", in *Proceedings of the USSR Academy of Sciences*. Technical Cybernetics, No.1, 1974 (in Russian).
53. P.S. Sapaty, "The WAVE-0 language as a framework of navigational structures for knowledge bases using semantic networks", in *Proceedings of USSR Academy of Sciences*. Technical Cybernetics, No. 5 (1986) (in Russian).
54. P.S. Sapaty, "The wave approach to distributed processing of graphs and networks", in *Proceedings of International Working Conference Knowledge and Vision Processing Systems*, Smolenice, November 1986.
55. P.S. Sapaty, A wave language for parallel processing of semantic networks. *Computers & Education: Artificial Intelligence*, 5(4) (1986), 289–314.
56. P. Sapaty, S. Varbanov, M. Dimitrova, "Information systems based on the wave navigation techniques and their implementation on parallel computers", in *Proceedings of International Working Conference on Knowledge and Vision Processing Systems*, Smolenice, November 1986.
57. P. Sapaty, I. Kocis, "A parallel network wave machine", in *Proceedings of 3rd International Workshop PARCELLA'86*, Akademie-Verlag, Berlin, 1986.
58. P. Sapaty, S. Varbanov, A. Iljenko, "The WAVEmodel and architecture for knowledge processing", in *Proceedings of Fourth International Conference on Artificial Intelligence and Information- Control Systems of Robots*, Smolenice, 1987.
59. P.S. Sapaty, "Distributed Artificial Brain for Collectively Behaving Mobile Robots", in *Proceedings of Symposium and Exhibition Unmanned Systems 2001*, Baltimore, MD, July 31–August 2 2001, 18 p.
60. P. Sapaty, A. Morozov, M. Sugisaka, "DEW in a Network Enabled Environment", in *Proceedings of the International Conference Directed Energy Weapons 2007*, Le Meridien Piccadilly, London, UK, 28 February–1 March 2007.
61. P.S. Sapaty, "Global Management of Distributed EW-Related System", in *Proceedings of International Conference Electronic Warfare: Operations & Systems*, 2007, Thistle.
62. P. Sapaty, M. Sugisaka, J. Filipe, "Making Sensor Networks Intelligent", in *Proceedings of the 4th International Conference on Informatics in Control, Automation and Robotics, ICINCO-2007*, Angers, France, 9–12 May 2007.
63. P.S. Sapaty, Crisis management with distributed processing technology. *International Transactions on Systems Science and Application*, 1(1) (2006), 81–92. ISSN 1751-1461.
64. P. Sapaty, M. Sugisaka, R. Finkelstein, J. Delgado-Frias, N. Mirenkov, "Emergent Societies: An Advanced IT Support of Crisis Relief Missions", in *Proceedings of Eleventh International Symposium on Artificial Life and Robotics (AROB 11th'06)*, Beppu, Japan, 23–26 January 2006, ISBN 4-9902880-0-9.
65. P.S. Sapaty, "Distributed Technology for Global Dominance", in *Proceedings of International Conference Defense Transformation and Net-Centric Systems*

2008, as part of the SPIE Defense and Security Symposium, World Center Marriott Resort and Convention Center, Orlando, FL, USA, 16–20 March 2008 (Proceedings of SPIE—Volume 6981, Defense Transformation and Net-Centric Systems 2008, Raja Suresh, Editor, 69810T, 3 (April, 2008).
66. P.S. Sapaty, "Distributed Technology for Global Dominance. Keynote lecture", in *Proceedings of the Fifth International Conference in Control, Automation and Robotics ICINCO 2008, The Conference Proceedings*, Funchal, Madeira, Portugal, 11–15 May, 2008.
67. P.S. Sapaty, "Human-Robotic Teaming: A Compromised Solution", in *AUVSI's Unmanned Systems North America 2008*, San Diego, USA, 10–12 June, 2008.
68. P. Sapaty, M. Sugisaka, M.J. Delgado-Frias, J. Filipe, N. Mirenkov, Intelligent management of distributed dynamic sensor networks. *Artificial Life and Robotics*, 12(1–2) (2008, March), 51–59.
69. P.S. Sapaty, "High-Level Communication Protocol for Dynamically Networked Battlefields", in *Proceedings of International Conference Tactical Communications 2009 (Situational Awareness and Operational Effectiveness in the Last Tactical Mile)*, One Whitehall Place, Whitehall Suite and Reception, London, UK (2009).
70. P.S. Sapaty, Distributed capability for battlespace dominance. Electronic warfare 2009 conference and exhibition, Novotel London West Hotel and Conference Center, London, 14–15 May, 2009.
71. P.S. Sapaty, Distributed capability for battlespace dominance. Electronic warfare 2009 conference and exhibition, Novotel London West Hotel and Conference Center, London (2009).
72. P.S. Sapaty, "Providing Spatial Integrity for Distributed Unmanned Systems", in *Proceedings of 6th International Conference in Control, Automation and Robotics ICINCO 2009*, Milan, Italy (2009).
73. P.S. Sapaty, *Distributed Technology for Global Control*. Book chapter, Lecture Notes in Electrical Engineering, Informatics in Control, Automation and Robotics, vol. 37 (Springer, Berlin, 2009).
74. P.S. Sapaty, Gestalt-Based Integrity of Distributed Networked Systems. SPIE Europe Security + Defence, bcc Berliner Congress Centre, Berlin, Germany (2009).
75. P.S. Sapaty, Remote Control of Open Groups of Remote Sensors. in *Proceedings of SPIE Europe Security + Defence*, Berlin, Germany (2009).
76. P.S. Sapaty, "Tactical Communications in Advanced Systems for Asymmetric Operations", *Proceedings of Tactical Communications 2010*, CCT Venues, Canary Wharf, London UK, 28–30 April 2010.
77. P.S. Sapaty, "High-Level Technology to Manage Distributed Robotized Systems", in *Proceedings of Military Robotics 2010*, Jolly St Ermins, London, UK, 25–27 May 2010.
78. P.S. Sapaty, Emerging Asymmetric Threats, Q&A Session. Tactical Communications 2010, CCT Venues, Canary Wharf, London UK, 28–30 April 2010.
79. P.S. Sapaty, "High-level Organisation and Management of Directed Energy Systems", in *Proceedings of Directed Energy Weapons 2010*, CCT, Canary Wharf, London UK, 25–26, March 2010.
80. P.S. Sapaty, "Formalizing Commander's Intent by Spatial Grasp Technology", in *Accepted Paper at the International society of Military Sciences (ISMS) 2012 Annual Conference*, Kingston, Ontario, Canada, 23–24 October 2012.

81. P.S. Sapaty, Unified Transition to Robotized Armies with Spatial Grasp Technology. International Summit Military Robotics, London, UK, 12–13 November 2012.
82. P.S. Sapaty, Distributed air and missile defense with spatial grasp technology. *Intelligent Systems, Control and Automation: Science and Engineering* 3(2) (2012), 117–131.
83. P.S. Sapaty, "Global Electronic Dominance", in *12th International Fighter Symposium*, 6th–8th November 2012, Grand Connaught Rooms, London, UK.
84. P.S. Sapaty, Providing global awareness in distributed dynamic environments. International summit ISR, London, 16–18 April 2013.
85. P.S. Sapaty, "Ruling distributed dynamic worlds with spatial grasp technology", in *Tutorial at the International Science and Information Conference 2013 (SAI)*, London, UK, 7–9 October 2013.
86. P.S. Sapaty, "Night Vision under Advanced Spatial Intelligence: A Key to Battlefield Dominance", in *International Summit Night Vision 2013*, London, 4–6 June 2013.
87. P.S. Sapaty, "Integration of ISR with advanced command and control for critical mission applications", in *SMi's ISR Conference*, Holiday Inn Regents Park, London, 7–8 April 2014.
88. P.S. Sapaty, Unified transition to cooperative unmanned systems under spatial grasp paradigm, *Transactions on Networks and Communications* 2(2) (2014, April).
89. P.S. Sapaty, "From Manned to Smart Unmanned Systems: A Unified Transition", in *SMi's Military Robotics*, Holiday Inn Regents Park London, 21–22 May 2014.
90. P.S. Sapaty, "Unified Transition to Cooperative Unmanned Systems under Spatial Grasp Paradigm", in *19th International Command and Control Research and Technology Symposium*, Alexandria, Virginia, 16–19 June 2014.
91. P.S. Sapaty, Distributed human terrain operations for solving national and international problems. *Int. Relat. Diplomacy* 2(9) (2014, September).
92. P.S. Sapaty, "Providing Over-operability of Advanced ISR Systems by a High-level Networking Technology", in *SMI's airborne ISR*, Holiday Inn Kensington Forum, London, UK, 26–27 October 2015.
93. P.S. Sapaty, "Distributed Missile Defense with Spatial Grasp Technology", *SMi's military space*, Holiday Inn Regents Park London, 4–5 March 2015.
94. P.S. Sapaty, Military Robotics: Latest Trends and Spatial Grasp Solutions, *International Journal of Advanced Research in Artificial Intelligence* 4(4) (2015).
95. P.S. Sapaty, Organization of Advanced ISR Systems by High-Level Networking Technology, MMC, No 1 (2016).
96. P.S. Sapaty, Towards Massively Robotized Systems under Spatial Grasp Technology, *Journal of Computer Science and Systems Biology*, 9(1) (2016).
97. P.S. Sapaty, Towards Wholeness and Integrity of Distributed Dynamic Systems, *Journal of Computer Science and Systems Biology*, 9(3) (2016).
98. P.S. Sapaty, Towards Global Goal Orientation, Robustness and Integrity of Distributed Dynamic Systems, *The Journal of Diplomacy and International Relations* 4(6) (2016, June).
99. P.S. Sapaty, Mosaic Warfare: From Philosophy to Model to Solution, *Mathematical Machines and Systems* 3 (2019).

100. P.S. Sapaty, *Symbiosis of Real and Simulated Worlds under Spatial Grasp Technology*. Springer, 2021. 305 p.
101. P.S. Sapaty, *Complexity in International Security: A Holistic Spatial Approach*. Emerald Publishing, 2019. 160 p.
102. P.S. Sapaty, *Holistic Analysis and Management of Distributed Social Systems*. Springer, 2018. 234 p.
103. P.S. Sapaty, *Managing Distributed Dynamic Systems with Spatial Grasp Technology*. Springer, 2017. 284 p.
104. P.S. Sapaty, *Ruling Distributed Dynamic Worlds*. New York: John Wiley & Sons, 2005. 255 p.
105. P.S. Sapaty, *Mobile Processing in Distributed and Open Environments*. New York: John Wiley & Sons, 1999. 410 p.
106. P.S. Sapaty, Global Network Management under Spatial Grasp Paradigm. *IRATJ* 6(3) (2020), 134–148. https://medcraveonline.com/IRATJ/IRATJ-06-00212.pdf
107. P. S. Sapaty, Global Network Management under Spatial Grasp Paradigm, *Global Journal of Researches in Engineering: J General Engineering* 20(5) (2020), 58–81. Version 1.0 Year 20. https://globaljournals.org/GJRE_Volume20/6-Global-Network-Management.pdf
108. P.S. Sapaty, Advanced Terrestrial and Celestial Missions under Spatial Grasp Technology, *Aeronautics and Aerospace Open Access Journal* 4(3) (2020). https://medcraveonline.com/AAOAJ/AAOAJ-04-00110.pdf
109. P.S. Sapaty, Spatial Management of Distributed Social Systems, *Journal of Computer Science Research* 2(3) (2020, July). https://ojs.bilpublishing.com/index.php/jcsr/article/view/2077/pdf
110. P.S. Sapaty, Towards Global Nanosystems Under High-level Networking Technology, *Acta Scientific Computer Sciences* 2(8) (2020). https://www.actascientific.com/ASCS/pdf/ASCS-02-0051.pdf
111. P.S. Sapaty, Symbiosis of Distributed Simulation and Control under Spatial Grasp Technology, *SSRG International Journal of Mobile Computing and Application (IJMCA)*, 7(2) (2020, May–August). http://www.internationaljournalssrg.org/IJMCA/2020/Volume7-Issue2/IJMCA-V7I2P101.pdf
112. P.S. Sapaty, Global Network Management under Spatial Grasp Paradigm, *International Robotics & Automation Journal*, 6(3) (2020). https://medcraveonline.com/IRATJ/IRATJ-06-00212.pdf
113. P.S. Sapaty, Global Network Management under Spatial Grasp Paradigm, *Global Journal of Researches in Engineering: J General Engineering* 20(5) (2020) Version 1.0. https://globaljournals.org/GJRE_Volume20/6-Global-Network-Management.pdf
114. P.S. Sapaty, Symbiosis of Virtual and Physical Worlds under Spatial Grasp Technology, *Journal of Computer Science & Systems Biology* 13(6) (2020). https://www.hilarispublisher.com/open-access/symbiosis-of-virtual-and-physical-worlds-under-spatial-grasp-technology.pdf
115. P.S. Sapaty, Symbiosis of Real and Simulated Worlds Under Global Awareness and Consciousness, The Science of Consciousness | TSC 2020. https://eagle.sbs.arizona.edu/sc/report_poster_detail.php?abs=3696
116. P.S. Sapaty, Spatial Grasp as a Model for Space-based Control and Management Systems, *Mathematical Machines and Systems* 1 (2021), 135–138. http://www.immsp.kiev.ua/publications/articles/2021/2021_1/Sapaty_book_1_2021.pdf

117. P.S. Sapaty, Managing multiple satellite architectures by spatial grasp technology, *Mathematical Machines and Systems* 1 (2021), 3–16. http://www.immsp.kiev.ua/publications/eng/2021_1/
118. P.S. Sapaty, Spatial Management of Large Constellations of Small Satellites, *Mathematical Machines and Systems* 2 (2021). http://www.immsp.kiev.ua/publications/articles/2021/2021_2/02_21_Sapaty.pdf
119. P.S. Sapaty, Global Management of Space Debris Removal under Spatial Grasp Technology, *Acta Scientific Computer Sciences* 3(7) 2021 July. https://www.actascientific.com/ASCS/pdf/ASCS-03-0135.pdf
120. P.S. Sapaty, Space Debris Removal under Spatial Grasp Technology, *Network and Communication Technologies*, 6(1) (2021). https://www.ccsenet.org/journal/index.php/nct/article/view/0/45486
121. P.S. Sapaty, Spatial Grasp Model for Management of Dynamic Distributed Systems, *Acta Scientific Computer Sciences*, 3(9), 2021 https://www.actascientific.com/ASCS/pdf/ASCS-03-0170.pdf
122. P.S. Sapaty, Spatial Grasp Model for Dynamic Distributed Systems, *Mathematical Machines and Systems*, 3 (2021). http://www.immsp.kiev.ua/publications/articles/2021/2021_3/03_21_Sapaty.pdf
123. P.S. Sapaty, Development of Space-based Distributed Systems under Spatial Grasp Technology, *Mathematical Machines and Systems*, 4 (2021).

9781032136097